Island Press ... d States whose principal purpose is the publication of books on environmental issues and natural resource management. We provide solutions-oriented information to professionals, public officials, business and community leaders, and concerned citizens who are shaping responses to environmental problems.

In 1994, Island Press celebrated its tenth anniversary as the leading provider of timely and practical books that take a multidisciplinary approach to critical environmental concerns. Our growing list of titles reflects our commitment to bringing the best of an expanding body of literature to the environmental community throughout North America and the world.

Support for Island Press is provided by Apple Computer, Inc., The Bullitt Foundation, The Geraldine R. Dodge Foundation, The Energy Foundation, The Ford Foundation, The W. Alton Jones Foundation, The Lyndhurst Foundation, The John D. and Catherine T. MacArthur Foundation, The Andrew W. Mellon Foundation, The Joyce Mertz-Gilmore Foundation, The National Fish and Wildlife Foundation, The Pew Charitable Trusts, The Pew Global Stewardship Initiative, The Rockefeller Philanthropic Collaborative, Inc., and individual donors.

More than ever, our effect on the biological systems of the planet will rebound to affect us. A slash and burn approach to the biosphere is no longer viable. Indeed the planet already has a reduced capacity to support Man. We need a populace and politicians aware that all decisions have a biological component, and that biology is inextricably interwoven with sociology and economics. As the planet becomes simpler biologically, it becomes more expensive economically: fish are smaller and dearer; lumber is narrower, shorter and more expensive; dwindling natural resources fuel inflation. The planet also is more vulnerable to disaster, and the quality of life inevitably declines.

Conservation is sometimes perceived as stopping everything cold, as holding whooping cranes in higher esteem than people. It is up to science to spread the understanding that the choice is not between wild places or people. Rather it is between a rich or an impoverished existence for Man.

Thomas E. Lovejoy

THE
WORK
OF
NATURE

THE
WORK
OF
NATURE

How the Diversity of Life Sustains Us

YVONNE BASKIN

A Project of SCOPE:
The Scientific Committee on Problems of the Environment

ISLAND PRESS
Washington, D.C. ● Covelo, California

First paperback edition published in 1998.

Library of Congress Cataloging-in-Publication Data

Baskin, Yvonne.
    The work of nature: how the diversity of life sustains us/ Yvonne Baskin.
        p.   cm.
    Includes bibliographical references and index.
    ISBN 1-55963-519-3 (cloth) — ISBN 1-55963-520-7 (pbk)
    1. Environmentalism. 2. Biological diversity. 3. Human ecology. 4. Conservation of natural resources.   I. Title.
GE195.B36   1997
333.7'2—dc21   96-52051
                                                                CIP

Printed on recycled, acid-free paper ✪

Manufactured in the United States of America

10 9 8 7 6 5 4 3

Contents

Foreword

Of Keystone Complexes and Nature's Services

Ecologists are well aware of the critical nature of the services supplied to humanity by natural ecosystems, but they sometimes are surprised at the subtlety and complexity of the interactions that can be involved in supplying them. For example, Tree Swallows and Violet-green Swallows help to provide an important ecosystem service in the vicinity of the Rocky Mountain Biological Laboratory (RMBL), where our group does summer research. That service is natural pest control; swallows snatch insects in mid-air, and thus greatly reduce the numbers of mosquitoes and other biting flies that otherwise make researchers' lives miserable. They also generally help to prevent outbreaks of insect populations. Recent work by Gretchen Daily of Stanford University and her colleagues illuminates some of the unexpected ways in which the swallows' role as pest-control agents depends on other species, including microbes.[1]

Around RMBL the swallows must nest in old woodpecker holes, the vast majority of which were made by Red-naped Sapsuckers. Sapsuckers are among the few non-human animals that don't simply consume resources, they work to mobilize them. Early in the season, sapsuckers drill holes in aspen and spruce trees, and then lap up the sugary sap that oozes from the holes, or "wells." They also excavate nest holes in the aspens. But when the young sapsuckers hatch, the adults switch to a new, rich source of sap: willow shrubs. They cut rectangular wells in the willow bark and peck away at the upper edge of the well to encourage the flow of sap, which averages about 29 percent sugar. Not only do the adults feed on the sap themselves; they dip captured insects in the high-energy fluid before feeding them to their nestlings. When the young are fledged, they join their parents at the wells.

There is one other major living piece to the puzzle. The sapsuckers can only excavate nest holes in aspen trees whose trunks have first been softened by infection with a heart-rot fungus. Thus the presence of the swallows depends upon a "keystone complex" consisting of sapsuckers, aspens, willows, and a fungus.

Furthermore, many other organisms—including wasps, butterflies, warblers, hummingbirds, chipmunks, and squirrels—steal sap from the sapsucker wells. In some areas, in the absence of flowers as nectar sources, the spring range of hummingbirds is extended by the presence of sapsucker wells. The warblers, hummers, and wasps also contribute to the pest-control service.

There are myriad other ways in which the delivery of ecosystem services depends on biodiversity—Earth's living wealth, the most important part of humanity's stock of "natural capital." Many are explained in *The Work of Nature.* It is a beautifully written volume, based on an extensive technical survey carried out by outstanding ecologists from all over the world. The book emphasizes that biodiversity is not simply an amenity, a luxury that humanity can do without. It explains that biodiversity is much more than simply the number of species that inhabits our planet, but that it includes critical genetic resources, a diversity of ecosystems and landscapes, and perhaps most crucial, a diversity of *populations.* The importance of the latter is demonstrated by a simple thought experiment. Suppose every species were reduced to just a single small population in a zoo, aquarium, botanic garden, or bucket of soil—the smallest population that would let that species persist for, say, one hundred years.[2] There would be no loss of species diversity, but all ecosystems, devoid of living organisms, would crash, and *Homo sapiens* would almost certainly go extinct.

The Work of Nature is an especially important book as the millennium approaches, when the American public is being bombarded with anti-environmental rhetoric and misinformation about biodiversity, denigrating the threat posed by the extinction crisis to the human enterprise.[3] Consider the following anti-environmental or "brownlash" statement:

> [East] of San Francisco [and] in all directions around Atlanta and
> Denver and Warsaw and Madrid, and in many similar locations
> worldwide, extensive tracts of habitat that have known only
> occasional human intervention abut centers of mechanistic
> human excess.[4]

But going east from San Francisco, one first finds highly modified and polluted San Francisco Bay, with the vast majority of its biodiversity-rich coastal marshes destroyed. Beyond the Bay are Oakland, Alameda, San Leandro, and other cities and suburbs, and then the completely agricultural Central Valley of California. In that area, usually bathed in a brown, human-created haze that can extend far into the Great Basin, virtually all natural habitat is gone. After the Central Valley come mountain ranges, where human intervention has exterminated the previously important top predator, the grizzly bear; completely changed the hydrological cycle; infested streams with giardia parasites; created vast damage by herding cattle and sheep, mining, timbering; and caused huge areas of the overgrazed intermountain valleys to be carpeted with introduced plants.

Beyond the vastly overpopulated front range of the Rockies, one reaches the Great Plains, where it is virtually impossible to find even small areas with a semblance of the original vegetation and minute remnants of the previously gigantic buffalo herds that once dominated them. Next come the intensively farmed prairies, where at best a few scraps of relatively unaltered grassland communities remain. In the eastern United States are farms and urban-suburban sprawl interspersed with fragmented and largely second-growth forests, often altered by acid rain and heavy-metal pollution. Wolves and cougars have been extirpated from the East, leaving a plague of deer and ticks that carry Lyme disease. Finally, one reaches the polluted and overfished Atlantic Ocean. Need I go on? In fact, as you will learn in the pages that follow, when human beings (perhaps aided by climate change) wiped out the Pleistocene megafauna, they altered the face of North America substantially and permanently. This was long before Europeans invaded the continent.

Contrary to the ill-informed assertion quoted above, the world is nearly devoid of habitats "that have known only occasional human intervention." Not only has most of the terrestrial surface been directly modified by building, paving, plowing, grazing, drilling, mining, clearing, logging, draining, pumping, or damming; but all of it has been affected by poisoning. Moreover, human intervention has more or less permanently altered the oceans by depleting fish and whale stocks, destroying coral reefs and coastal marshes, and emitting toxic pollutants. Indeed, every cubic centimeter of the biosphere has been altered by human-induced changes in the climate and the chemical composition of the atmosphere.

Purveyors of anti-environmental rhetoric apparently fail to appreciate the stakes in the game they are playing. The very future of humanity is in the balance. The scale of the human enterprise—the numbers of people multiplied by the average environmental impact of each—is now roughly twenty times greater than it was in 1850,[5] and as a consequence humanity has become a truly global force. Our species is altering the surface and atmosphere of Earth in ways unprecedented since the catastrophe that wiped out the dinosaurs 65 million years ago.

The most irreversible of human assaults on the environment is the one on biodiversity, for once extinct in the wild, populations usually can only be reestablished with great difficulty, and extinct species are gone forever. Humanity depends entirely on ecosystem services, and as this fine volume demonstrates, to a large degree those services depend on biodiversity. In essence, by killing off its only known living companions in the universe, *Homo sapiens* is destroying its own life support systems and catapulting itself toward ecological disaster. I hope *The Work of Nature* will help stop that process before it is too late.

Paul R. Ehrlich
Stanford University

Preface

Commissioned by the Scientific Committee on Problems of the Environment (SCOPE), this book summarizes and also marks the culmination of an exciting and innovative scientific endeavor. The goal of that endeavor was to answer a single key question: What are the possible consequences of the accelerating losses in biodiversity? Until recently, major concerns regarding biological extinction have been mostly ethical ones involving questions about our responsibility for the earth's biological heritage, and economic ones focusing on the potential loss of such economically valuable products as drugs, herbs, and foodstuffs.

The Work of Nature looks at biological diversity differently. Rather than examining ecosystems in terms of the "goods" (food, fiber, etc.) they provide, the book assesses biodiversity in terms of the earth's ability to provide the ecological "services," such as clean air and water, upon which humankind depends. Indeed, the major question of the book is, Are these services being impaired as we lose biological diversity?

The search for answers began in 1991, when SCOPE launched a program specifically to review and evaluate all scientific literature relevant to the subject of ecosystem functioning. We, as chair of the SCOPE Scientific Steering Committee responsible for the program and member of the Executive Committee, respectively, felt such a program was well suited to SCOPE's overall agenda. Formed by the International Council of Scientific Unions and based in Paris, SCOPE is a nongovernmental body, which has thirty-eight member nations and access to some of the world's best scientists. Indeed, SCOPE regularly calls on scientists from around the world, asking them to provide assessments of crucial global environmental problems. The organization then serves as a powerful vehicle for translating scientific knowledge into a format that is both accessible and timely. Among the issues SCOPE has tackled are the environmental consequences of nuclear conflict, climate warming, and altered global biogeochemical cycles.

The SCOPE Program on Ecosystem Functioning of Biodiversity mobilized hundreds of scientists around the world, specialists in the major ecosystems of the earth, including coral reefs, tropical forests, deserts, and tundra. Coming together initially in a series of ecosystem-specific workshops, these scientists reviewed existing data on the impact of diversity losses on ecosystem functioning. Thereafter, they met as a single group to compare results from their respective systems and assess similarities and differences. We are grateful to the John D. and Catherine T. MacArthur Foundation for supporting this phase of the study.

Subsequently, the United Nations Environment Program (UNEP) launched a Global Biodiversity Assessment, which incorporated and expanded the SCOPE project. Overall, the assessment was ground breaking because it represented the first time the significance of biodiversity loss on ecosystem functioning had ever been addressed.

By pushing ecological study in an entirely new direction, certain challenges were created, among them the need to evaluate evidence collected for other purposes. No direct experimental studies of the impact of biological loss on ecological services had ever been undertaken. Thus, much of the data that was needed to shed light on the question had to be drawn from interpretations of natural patterns and the ecological responses to both additions of species, resulting from invasions or introductions, and deletions of species, caused, for example, by selective harvesting. The SCOPE project relied on this data to draw the conclusions that form the foundation for this book.

Believing the topic to be both vitally important and timely, we encouraged the SCOPE Scientific Steering Committee to commission a book that would effectively deliver the project's findings to a broad audience. The result is the book before you. Indeed, *The Work of Nature* represents an enormously successful effort to translate scientific documents into a form accessible to a general reader. Yvonne Baskin has carried out this translation in a very special way. In a relatively short time, she not only mastered the significance of vast amounts of highly technical information but she also sought out and talked with many of the key scientists involved in the research she describes. Her words explain with exceptional clarity and eloquence how scientists deeply involved with these crucial issues think about their subject. We thank you, Yvonne.

Harold A. Mooney
Paul S. Achilles Professor of
Environmental Biology, Stanford
University

Jane Lubchenco
Wayne and Gladys Valley Professor of
Marine Biology, Oregon State
University; Oregon State University
Distinguished Professor; and
President, American Association for
the Advancement of Science, 1996

Acknowledgments

My first encounter with SCOPE and its project on ecosystem functioning and biodiversity took place in February 1994 when I was sent to
Asilomar, California, to write a news article about the group's findings
for *Science* magazine. Those five days at a SCOPE workshop amid the
Monterey pine groves on the Pacific coast provided me, a science jour-
nalist, with an entirely new perspective on the natural world and the
vital roles species play in generating our ecological life-support systems.
It was the beginning of a process of discovery that necessarily shifted
into warp speed a year later when Harold A. (Hal) Mooney asked me to
convert the group's three-year effort into a book for general readers.
First and foremost, then, I am fundamentally indebted to the hundreds
of scientists around the world who took part in the SCOPE project. Without their syntheses, recently published in half a dozen technical volumes, this book could not have been written.

I was especially fortunate to have the guidance of the scientific advi-
sory committee assembled for this project: Hal Mooney of Stanford University, Jane Lubchenco of Oregon State University, Anthony C. Janetos
of NASA's Mission to Planet Earth, Osvaldo E. Sala of the University of
Buenos Aires, Rodolfo Dirzo of Universidad Nacional Autonoma de Mexico, and J. Hall Cushman of Sonoma State University. Their careful read-

ing of the first draft of the manuscript provided me with invaluable criticisms and new insights that helped to improve the final version. I'm particularly grateful to Hal and Jane, who repeatedly and generously provided assistance, advice, encouragement, and enthusiasm during early fits and starts as the book took shape and who later conscientiously read and critiqued multiple drafts.

I also owe special thanks to a number of reviewers who read one or more chapters and offered helpful suggestions and tactful advice, including Edith B. Allen of the University of California, Riverside; Gary Allison at Oregon State University; Stephen Carpenter of the University of Wisconsin, Madison; John J. (Jack) Ewel of the Institute of Pacific Islands Forestry in Honolulu; Christopher Field of the Carnegie Institution of Washington; Diana Freckman of Colorado State University; Lawrence E. Gilbert of the University of Texas, Austin; Elaine Ingham of Oregon State University; Simon Levin of Princeton University; John Pastor of the University of Minnesota, Duluth; Louis F. Pitelka of the Appalachian Environmental Laboratory in Frostburg, Maryland; William Schlesinger of Duke University; David Tilman of the University of Minnesota, St. Paul; and Joy Zedler of San Diego State University. In addition, I'm deeply grateful to Nancy Huntly of Idaho State University, who read the entire manuscript and provided a very encouraging appraisal along with her detailed and thoughtful critique.

In the end, of course, it remains the author's task to correct and revise the manuscript, and any errors or deficiencies that remain after so much good advice are her responsibility alone.

I want to express deep appreciation also to Laurie Burnham, editorial director of Shearwater Books at Island Press, who devoted an enormous amount of time, thought, and much-needed advice to this project, starting long before this book was officially her responsibility. My thanks also to SCOPE for providing financial support for this project, and particularly to SCOPE Executive Director Veronique Plocq-Fichelet, who handled the contractual details of getting the book into print.

Most of all, I'm thankful for the terrific personal and professional support of my husband, Mike Gilpin, who kept me on course, guided me out of deadends, and helped lighten my inevitable black moods when it seemed the book would never be finished. He was the first to see each chapter in its roughest form and always provided invaluable direction while helping to redeem my ecological naivete. All this while enduring an acute loss of attention and companionship for months on end. Finally, my parents, though far away, were always a steadying and supportive presence, in part because the love of nature and the outdoors they instilled in me made the inevitable pain of producing this book seem worthwhile.

CHAPTER ONE

This Web of Life

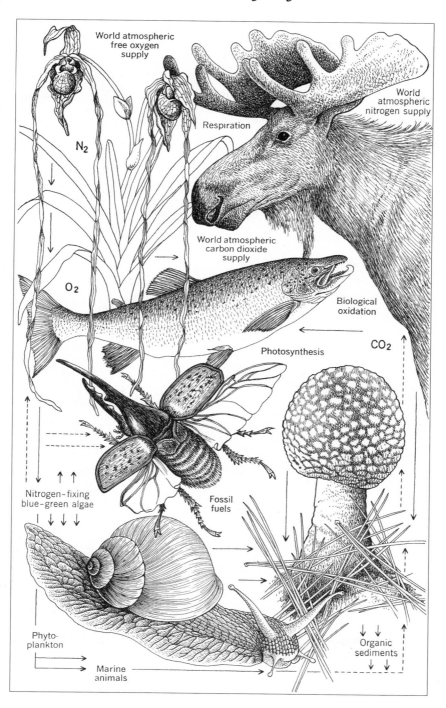

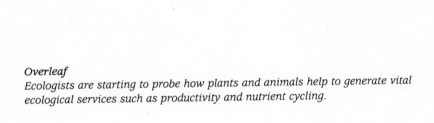

Overleaf
Ecologists are starting to probe how plants and animals help to generate vital ecological services such as productivity and nutrient cycling.

*O*ur oldest faiths and deepest symbols reflect a primal connection to the natural world, to a living planet that long ago imprinted on the human consciousness a cyclic sense of death and decay, rebirth and renewal. We do not question that flesh and bone and leaf litter will decay to dust, that seeds will sprout season after season and find renewed nourishment in the soil, that rivers can flow endlessly without running dry, that we can breathe for a lifetime without depleting the air of oxygen. Despite our fascination with other worlds and our hopeful probing of outer space, we've found no other planet where any of these things are true. What humans have not fully appreciated until recently is that these services are the work of nature, performed by the rich diversity of microbes, plants, and animals on the earth.

It is this lavish array of organisms that we call "biodiversity," an intricately linked web of living things whose activities work in concert to make the earth a uniquely habitable planet. But today, as never before, the species in this web are under siege, threatened by human activities that encroach on their habitats. At the same time, ecologists are increasingly aware that the impoverishment of species—the planet's work force—threatens to erode the basic life-support services that render the earth hospitable for humanity. Indeed, we are approaching a crossroads in time, when the survival and extinction of other species may well delimit the future of *Homo sapiens*.

Consider what life has done and continues to do for the earth.

Some 4 billion years ago, the primordial atmosphere was a ghastly brew, devoid of oxygen and unable to shield the earth's surface from the scorching, molecule-cleaving ultraviolet radiation of the young sun. Eventually life changed all that. Over billions of years, photosynthetic organisms in the sea released enough oxygen to create a protective ozone shield and a reservoir of free oxygen that allowed the first plants to venture onto the land. Through the alchemy of enzymes and solar energy, green plants from plankton to redwoods still carry on photosynthesis, turning water and carbon dioxide into free oxygen and also the carbon-based sugars needed to build all living tissues. These are the raw materials that underpin the earth's food webs and generate the food, fiber, timber, and fuel that sustain human societies.

Together, plants, animals, and microbes perform an array of vital services. They generate and preserve fertile soils. They break down organic wastes, from leaf litter to feces and flesh, recycling the mineral nutrients, carbon, and nitrogen needed for new plant growth. They absorb and break down pollutants; help maintain a benign mix of gases in the atmosphere; regulate the amount of solar energy the earth absorbs; moderate regional weather and rainfall; modulate the water cycle, minimizing floods and drought and purifying waters; blunt the impact of the seas that batter the land margins; pollinate crops; and control the vast majority of potential crop pests and carriers of human disease.

In addition, this rich abundance of organisms serves as a "genetic library," a catalog of solutions to the problems of living on the earth. This catalog is written in the language of DNA, and from it human societies have derived crops, livestock, medicines, and many other commodities.[1]

On a larger scale, the earth's various species form populations that are aligned into communities and ecological systems—ecosystems, for short—which deliver such subsidies as clean air, pure water, and lush landscapes. Ecosystems, flexibly defined, are living communities interacting with the physical environment in a specific geographical place. An ecosystem may be as small as a rotting log or a pond or as large as a spruce forest or a vast lake. Of course, species are the critical components, the cogs and wheels of functioning ecosystems. Lose too many species from a forest—the trees, the truffle-forming fungi on their roots, the insects that prey on tree-destroying pests, the beavers that create ponds and meadows amid the woods—and at some point the assemblage ceases to work like a forest.

One question increasingly on the minds of ecologists is, how many species can the earth's communities lose before the ecological systems that nurture life begin to falter? To take the extreme, if nature had to run with a skeleton crew, what organisms would be absolutely vital to maintain the earth as a living planet? Probably the only truly indispensable groups of organisms are the plants that capture carbon and solar energy and the ranks of decomposers that release the nutrients and energy in dead plant litter for reuse.[2] But a conservation agenda based on this extreme would ignore the elaborate tangle of loops and flourishes in the

food webs, the intricate array of consumers that eat plants and predators that eat consumers, the symbionts, parasites, and other hangers-on
who have claimed places for themselves in the earth's myriad communities. Such an agenda would overlook virtually all the charismatic creatures on today's conservation hot lists: pandas, wolves, elephants, bald
eagles. Unfortunately, that agenda would also sacrifice civilization,
which is supported by many of those loops and flourishes in the web of
life.

If we are realistic about our dreams for tomorrow, our goal is not really "saving the planet" in some minimalist form, but perpetuating its atmosphere, climate, landscapes, and living services in a state that allows
human civilizations to prosper. For that to occur, we need to preserve
natural systems that are rich, healthy, and resilient enough to continue
to support human welfare and economic activity for the next decade,
the next century, and beyond. Some twenty-five years ago, as the space-
age metaphor of "Spaceship Earth" took hold, ecologists Eugene P.
Odum of the University of Georgia and his brother, Howard T. Odum, of
the University of Florida first used the engineering term "life-support
systems" to describe the earth's self-renewing, life-giving natural ecosystems.[3,4] It is these systems, not a mere skeleton crew, that human societies must seek to maintain.

Thus, the real questions facing ecologists today center on how much
biological diversity any particular ecosystem needs to remain functional,
self-sustaining, and life supporting. How many species must humanity
protect, and which ones and where, to assure pure water supplies from
an alpine watershed; to preserve the fertility of tropical soils; to prevent
cactus and shrubs from taking over productive grasslands; to maintain
local rainfall patterns; to nurture coastal shrimp and fish populations; or
to assure the integrity of pristine wildlands we value for recreation,
tourism, or cultural traditions?

Conservationists have long asserted that every species counts to some
degree in keeping the earth's life-support systems working. Some play
crucial roles day to day; others step into the breach only in times of
stress or disturbance. Until recently, however, researchers have made little concerted effort to define the functional roles played by specific
plants, animals, or microbes. In truth, function itself is a human concept.
Organisms were not designed by natural selection to fill slots on an assembly line; each organism strives to make a living and reproduce itself.
But as it eats, grows, excretes waste, and moves about, disturbing the

physical environment, it unwittingly plays a part in generating grander processes that alter the flow of water, the recycling of energy and materials, the renewal of the atmosphere.

An Emerging Science

It may seem surprising that scientists know so little about species' roles in generating ecological services. Yet unraveling links between the feeding behavior of an animal or the nutrient-cycling traits of a plant and subsequent changes in the character of a landscape or the chemistry of the soil may require years of observations, experiments, and analysis. Sometimes this work spans professional lifetimes and thus extends well beyond the practical scope of graduate student projects and the longevity of most research grants.

The questions also cross the lines of professional disciplines. Until recently, there were specialists who studied ecosystems and others who studied species and populations, and there was minimal discourse between them. Ecologists concerned with ecosystem processes such as nutrient cycles and energy flows, for instance, have traditionally focused on quantifying the work that gets done in a given system, not who does it. These ecologists might chart the flow of energy through a food web, filling in numbers atop arrows that run from plant producers to herbivore consumers, predators, and, finally, decomposers. Few have attempted to fill in the names and numbers of the organisms doing the work at each level, much less to determine how many of the species could be lost before the cycles faltered and the arrows had to be erased.

Population biologists, for their part, ask what determines the abundance of plants and animals, why certain creatures live where they do, how they interact with their neighbors, and why one species is part of the community while another is not. They also look at food webs and study herbivory and predation, but traditionally they have not translated into quantified flows of carbon the dynamics, say, of moose browsing on birches or wolves feasting on elk.

Scientists now recognize, however, that the two research traditions must come together if they are to understand how biological losses impair ecological services. As part of a broader movement to integrate these traditions, the Scientific Committee on Problems of the Environment (SCOPE) launched a project in 1991 that would lay the foundation for an emerging scientific discipline linking ecosystem processes with biological diversity.

SCOPE is a committee of the International Council of Scientific Unions (ICSU), a nongovernmental body established in 1931 to promote international science and its application for the benefit of humanity. SCOPE was created by ICSU in 1969 to identify gaps in environmental knowledge, assess research needs, and provide syntheses of existing scientific information on emerging environmental issues. Its projects are funded by a wide variety of sources, from United Nations agencies, the World Bank, and the European Communities to various governmental agencies and private foundations. The projects are carried out by scientists who donate their time and expertise.

During the three-year SCOPE program on ecosystem functioning, hundreds of scientists from around the world came together at a series of meetings and workshops to synthesize what is known so far about the functional role of biodiversity at all levels. Biodiversity is most often talked about in terms of numbers of species, yet the SCOPE teams expanded that perspective to include the complexity, richness, and abundance of nature at all levels, from the genes carried by each local population of a species to the layout of communities and ecosystems across the landscape.

Although much of the information came from the literature of ecology, the SCOPE team also took a much broader view, drawing on relevant studies from a wide array of other sciences: forestry, soil science, agriculture, plant pathology, range management, fisheries, microbiology, limnology, oceanography, marine biology, hydrology, meteorology, and atmospheric sciences. Results of the team's synthesis have been published in a number of technical volumes, many of them dealing with specific biotic regions of the world such as savannas, tropical forests, islands, and arctic and alpine ecosystems.[5]

The Work of Nature relies heavily on the ideas and examples assembled during the SCOPE project. It is intended to introduce students, policymakers, and concerned citizens to the risks our own species incurs when we impoverish the rest of the life of the planet. This book is not meant to be an exhaustive account of species' roles in providing life-support services but rather a suggestive and cautionary tale from the frontiers of a new science. The SCOPE assessment showed that most of the information scientists have accumulated so far about ecosystem functioning and biodiversity deals with species rather than diversity at other

levels, from genes to landscapes, and so this book focuses predomi-
nantly on the work of species.

The SCOPE assessment makes it clear that species are not ecologically
equal; some are more valuable in terms of service rendered than others.
Yet so little is known about the prodigious array of organisms on the
earth, and so few names have been entered into the flow charts of energy and materials, that it would be both premature and irresponsible to
declare any species, however humble, expendable.

How should scientists begin, then, to identify those species and natural communities whose loss will cause the greatest impacts on ecological processes? One approach the SCOPE teams took was to look for
clues indicating which species play indispensable roles—either as dominant species or as so-called keystones, which are less imposing creatures from a human point of view, but ones that hold the key to the integrity of their communities. *The Work of Nature* explores the progress
ecologists have made so far in learning how to spot such creatures, and
also in determining whether most communities have alternates—
backup workers—that can step in and compensate for the loss of a dominant or keystone species.

Another approach taken by the SCOPE researchers was to examine
ecosystems directly to determine how the traits or activities of individual species might influence ecological processes. *The Work of Nature*
highlights examples where shifts in species have affected

- the persistence and stability of natural communities;
- water quality and flow and the health of aquatic habitats;
- soil fertility and its relation to healthy crops and forests;
- the productivity or lushness of both our wild lands and agricultural
  lands;
- the look and functioning of the landscape and the frequency of disturbances, such as fires; and
- local rainfall patterns and weather, as well as the state of the atmosphere and global climate.

The Biodiversity Crisis

The Work of Nature, like the emerging science it portrays, comes at a critical time. We are in the midst of a biodiversity crisis, an ongoing epi-

demic of species losses that British conservation scientist Norman Myers has labeled "biodepletion."[6] The richness and complexity of the natural world is declining at an ever-accelerating rate, as the earth's burgeoning human population strives for a steadily rising technological standard of living. Natural diversity is being brutally simplified to make way for a dizzying blend of artificial landscapes—villages, housing developments, parking lots, roads, factories, mines, shopping malls, schools, parks, gardens, golf courses, plantations, and croplands.

The biggest threats to the diversity of life on the earth are habitat loss, introduction of alien species into communities, and fragmentation of natural areas caused by bulldozing, paving, plowing, draining, dredging, trawling, dynamiting, and damming. Humans are also plundering natural communities by overharvesting, overgrazing, dousing them with excessive pesticides and herbicides, raining acids and other pollutants onto them, altering the mix of gases in the air, and even thinning the ultraviolet radiation shield on which terrestrial life depends.[7]

Many of these assaults are so massive they wipe out entire ecosystems and disrupt natural processes immediately and directly. For example, draining and filling wetlands or permanently stripping the forest from a watershed instantly eliminates the flood and erosion control, water filtration and purification, and other services those ecosystems provide. Dynamiting a coral reef to extract fish not only destroys the ecosystem, but also exposes the now-unprotected shoreline to storms and so threatens coastal habitats. The impacts of such obvious forms of destruction are immediate and direct. Of equal concern to many scientists, however, is the slower, usually more insidious chipping away of functioning that accompanies the loss of species and impoverishment of their habitats. This erosion of service is harder to spot until it's well underway—easier for developers and government officials and the public to ignore for the moment. After all, what's so serious about losing a few more hectares of land to a few more houses?

Yet even simple questions, such as, "how quickly is this subtle impoverishment proceeding?," have no easy answers. Furthermore, the answer is endlessly contentious, partly because scientists have no clear tally, even to the nearest tens of millions, how many species share the earth with us. For instance, about 1.4 million species of plants, animals, and microbes have been formally identified and given Latinized names. But that itself is an estimate because no computerized database of the life of the planet has ever been compiled—a stunning omission given the great care and expense with which every newly discovered bit of the human genetic code is digitized and celebrated into cyberspace.

These named species are but a fraction of the actual diversity of life on the planet. Scientists know that because every time a trained researcher explores some poorly studied bit of habitat—the floor of the deep sea, the canopy of a tropical rain forest, a scoop of ordinary soil—he or she comes upon a wealth of previously unknown species. In one classic case, Terry L. Erwin of the Smithsonian Institution in Washington, D.C., misted the canopy of rain forest trees in Panama and elsewhere with an insecticidal fog and found that up to twelve hundred beetle species, many new to science, tumbled into his nets from a single species of tree. Many of the beetles were specific to a single tree species, existing nowhere else. Extrapolating from these counts, using the estimated number of tree species in the tropical forests, Erwin calculated that there may be 30 million species of insects and other arthropods in the tropical forests of the world.

The soil with its indispensable work force of specialists in rot, decay, and renewal may boast another 2 to 3 million species of bacteria and 1.5 million species of fungi, 95 percent of them still unknown. The largely unprobed seas undoubtedly contain millions more microbes. Although 99 percent of the species on earth are believed to fall into the smaller-than-a-bumblebee category, every year two to three new bird species are found, and now and then a new mammal.

All these factors have led biologists Paul Ehrlich of Stanford University and Edward O. Wilson of Harvard University to estimate that the diversity of life on earth may range as high as 100 million species.[8,9] The Global Biodiversity Assessment released in 1995 by the United Nations Environment Program (UNEP) expressed a more conservative consensus, giving a range for the total number of species between 7 and 20 million and suggesting 13 to 14 million as a "good working estimate."[10]

Basing their estimates on the fact that the vast majority of species have limited distributions (as evidenced by Erwin's study as well as that of others), Ehrlich and Wilson speculate that one quarter or more of the species on earth could be eliminated in the next fifty years if current rates of deforestation continue in the tropics, where more than half of the earth's species are thought to live. During that same time period, and not by harmless coincidence, the human population will double to more than 10 billion.[11]

Nay-sayers argue that estimates of species numbers are too vague and that, in any case, change and extinction are only natural. It's true that natural communities are not static. The "balance of nature" is a dynamic one. New species arise, others go extinct, and the assemblage regroups after ice ages, volcanic eruptions, or other upheavals. What is different about today's human-driven extinction threat, however, is its pace and scale.

Extinction rates today exceed by one hundred to one thousand times those seen in the fossil record. Ecologists point out that these rates will be ten times higher if all the species now officially listed as threatened or endangered actually disappear in the next century. With the human population growing exponentially, demanding more land, more food, more resources, millions of species may go extinct before they can be identified and their importance determined.[12]

Skeptics may still argue that massive losses, mostly among obscure species in localized areas of the tropics and other "hot spots" of high diversity, will have little impact on the human enterprise, especially in temperate regions. Yet the complete extinction of species is only one aspect of the biodiversity crisis. A more urgent but lesser known problem already eroding the structure of communities and the provision of ecological services is the dwindling of numerous plant and animal populations.

Ehrlich and Gretchen Daily, also of Stanford University, point out that a species cannot go extinct until all of its populations throughout its full territorial range have been lost. Yet long before the last remaining individuals die, the functional or economic value of the species fades. Few people can make a living off the meager populations of great whales, bison, or Pacific sardines that survive today after rampant overharvesting brought them to the brink of extinction. Ten sockeye salmon struggling up the Snake River into Idaho don't serve bears, bald eagles, or fishermen the way 10 million once did. Certainly, by the time most creatures have been officially classified as threatened or endangered, their numbers are too low to have much impact on ecosystem services across the landscape. Besides, organisms can only work where they live.[13,14]

If the red mangroves of Florida were wiped out, the coastal fisheries they nourish would suffer despite the persistence of similar mangrove forests along the Pacific Coast of South America. Eucalyptus trees have been transported to every continent for timber and horticulture, yet

When species such as bison dwindle in number, their ecological value fades long before they reach the brink of extinction.

that's no comfort in the wheat-growing regions of Australia, where their loss has led to waterlogging and salt damage of soils. Elephants thriving in a southern African game park do nothing to prevent bush encroachment on the East African savannas. It goes without saying that animals and plants reduced to surviving in zoos and botanic gardens or preserved as tissue specimens in "frozen zoos" or seed banks cannot play a role in ecosystem functioning. To do so they must be restored in healthy numbers to their native communities. Assuming, of course that those communities are fit for them to return to.

Biodiversity is not just a numbers game. In places like Hawaii and Florida, human introductions of new species, both deliberate and accidental, have actually increased the absolute count of species. But this "enrichment" has come at the cost of impoverishing the native species and sometimes altering the functioning of natural ecosystems. Throughout the world, exotic species are invading or being introduced into natural communities at an alarming rate, threatening to homogenize the global landscape to the point that Gordon Orians of the University of Washington has suggested calling the coming era the "Homogocene."

A pressing question is how well weedy or impoverished systems can maintain the life-support functions on which we rely. Too often, when humans disturb or deliberately "simplify" a landscape—say, clearcut forests or turn prairies to monocultures of corn—the result is a "leakier" system that lets more energy, nutrients, and topsoil slip away and shows less resistance to pests and other natural shocks. These aren't the robust systems on which most of us would bank our futures.

Stewardship and Self-Interest

Most of us give little thought to this impoverishment because we are increasingly estranged from the workings of the natural world. Nearly half of us live in urban settings, and that figure is climbing. North America and Europe were more than half urban by mid-century, and the shift of people from the countryside into the cities is increasing across Asia, Africa, and Latin America. By 2025, some 60 percent of the world's population will live in cities, increasingly remote from the natural communities that serve us.[15] We humans traditionally have trouble grappling with issues that we cannot see firsthand. Cloistered in our homes and offices, moated away from wildness by clipped lawns and pavement, nourished on piped water and shrink-wrapped foods, it's easy to lose sight of our reliance on plants, animals, insects, and microbes, as well as the cyclical processes they drive.

Nature in a city can too quickly be reduced to ranks of imported trees lining our boulevards, pets, exotic zoo animals, and gardens of wondrous flowers assembled at whim into unlikely "communities" plucked from every continent—African daisies nestling at the foot of South American bougainvilleas and Australian eucalyptus. For all the romantic appeal or sensory delight these living things provide, it's hard to conjure up a deep sense of dependence on such picturesque bits of nature, artificially nurtured in settings of our choice.

Until recently, the functional role of organisms has seldom been invoked as a basis for conservation. Indeed, ethical and moral pleas for saving species still predominate, bolstered by spiritual traditions in most cultures that include some moral embrace of the life of the planet. At its best, the conservation ethic in the Western world is based on a sense of stewardship, an obligation to protect and care for our only known companions in the universe. Yet for all its noble intent, that moral commitment has proven a slippery foundation for conservation. We continue to degrade and impoverish biological communities at an unprecedented pace.

The reasons for such folly are complex, ranging from economic desperation to simple ignorance or even sheer unconcern about the impacts of our actions. Human societies have a sad history of setting moral burdens aside while acquiring more comfortable or prosperous lifestyles.

Because of this, some ecologists and conservationists have been heartened as the economic benefits of biodiversity have started to become apparent. In recent decades, the direct economic value of the natural world as a source of everything from antibiotics and novel medicines to Brazil nuts, salmon, spices, mushrooms, mahogany, oils, and ecotourist dollars has been touted.

Like the service function of biodiversity, however, this growing emphasis on the practical value of species and natural systems is disquieting to many who harbor a deep affection for the natural world. To some, it seems superfluous on one hand that a creature as marvelous as a moose might have to pin its survival on human self-interest. On the other hand, it hardly seems demeaning to recognize that a moose strongly shapes the character of the very soils and trees in its forest. Such knowledge is likely to raise the general level of respect for less charismatic organisms in the moose's ecosystem, such as fungi, termites, and plankton, which have few advocates, even among traditional conservationists.

The key to self-preservation lies in understanding how species contribute to the functioning of ecosystems and how the forces that threaten biodiversity may alter vital ecological services. *The Work of Nature* describes the beginnings of this quest for understanding, offering not only an overview of how ecosystems operate but also a glimpse of the kinds of experiments that may illuminate the crucial role our fellow creatures play in sustaining the living earth.

CHAPTER TWO

The Keystone Club: Who's Important

Overleaf
Sea otters are keystone predators, devouring enough sea urchins to prevent these spiny grazers from devastating the kelp forest.

*I*t was a small and exclusive club at first. Predators only. Now it's hard to predict who might show up next in this unlikely assemblage of elites. Snails in the Negev Desert of Israel. Fig trees in the rain forests of Amazonia. Elephants on the African savanna. Termites, lobsters, beavers, and tsetse flies. All of them have been nominated as "keystone" species. Like a keystone in an arch—the wedge-shaped stone at the pinnacle that stabilizes the entire span—populations of each of these organisms seem to wield a disproportionate power over the structure of their communities, much more than size or numbers would suggest.

Membership in the keystone club was once bestowed on a select few creatures at the top of the food web, predators whose appetites not only controlled the abundance of prey species but indirectly influenced the mix of other organisms within the community. Ecologists now regard the keystone club as a much more diverse organization, and newer nominees may exert their power in any number of ways, from controlling the supply of key resources to altering the flow of water across the landscape.[1]

The expanding definition of keystone species reflects a growing effort on the part of ecologists to identify which organisms are most important to the health and maintenance of ecological processes that comprise the earth's life-support systems. The integrity of processes at the ecosystem level, of course, depends on the integrity of communities—the assemblage of plants, animals, and microbes that live and interact in a particular habitat. Scientists must know how natural communities are organized; they must know how the tug and pull between species affects the structure and stability of these communities if they are to predict when and where losses or additions of species will cause system-level processes to falter.[2]

A fundamental question underlies this effort: Is every species important to the integrity of the planet's life support services? There are two extremes on the spectrum of possible answers. One is the view that each species is unique and important, and its loss will have demonstrable consequences for the community. The other is the view that species are broadly interchangeable, meaning that most extinctions will cause little or no loss of ecosystem functioning because survivors can fill in for them. No credible ecologist espouses either of these extremes.

Some species clearly have more impact on their community than others. The most abundant species, the ones that dominate the space and resources in a community or define its character, usually also contribute most to controlling the ecosystem's productivity, nutrient cycling, and other processes. Obviously, redwood trees define a redwood forest, reef-building corals a coral reef community, kelp a kelp bed, certain grasses a prairie. Removing a coral reef will destroy the habitat for fish and other creatures, modify water currents, and expose coastal seagrass beds and mangroves to the full force of storm waves. Cutting down a forest or planting a new type of dominant tree can change the chemistry of the soil; the way water and nutrients are cycled; the amount of sunlight reaching the ground; the nature of the organisms that can exist in the understory; the productivity of forest streams; even local, and sometimes regional, rainfall and temperature.

Elimination of one of these abundant or dominant species should have a much larger impact on how an ecosystem functions than removal of most rare species. Unless, of course, that rare species has unique traits that make it a keystone or allow it to step in and do the work of other species in times of stress or disturbance. Communities may harbor hundreds, even thousands, of species that individually appear to be rather powerless. Yet sorting out their true ecological influence has proven a complex task.

In addition to the keystone concept, two other hypotheses have been proposed to characterize the relative influence of species in communities and therefore the likely ecological consequences of species losses. These are widely known as the rivet and the redundant species hypotheses.

Ecologists Paul and Anne Ehrlich of Stanford University in 1981 set the philosophical stage for current discussions of the importance of species when they proposed an analogy that became the rivet hypothesis. They compared the richness of species in an ecosystem to the rivets holding an airplane together. They noted that popping rivets out at random undoubtedly weakens an airplane and hastens the day when some threshold is reached and the plane breaks down or crashes. Over the years, this hypothesis has evolved to incorporate the notion that some rivets, such as keystone species, are more critical—or more strategically placed—than others, and their loss would cause more immediate harm to the plane. The analogy does not mean that every popped rivet, or every lost species, compromises the workings of the system. Indeed, scientists know that natural systems have some capacity to buffer losses because some background level of extinction occurs naturally and continually. The need for great caution in consigning species to extinction stems from our ignorance of how many rivets, and which ones, any particular community can afford to lose without compromising higher-level ecological services.[3] As the Ehrlichs stated in 1981:

> Ecosystems, like well-made airplanes, tend to have redundant subsystems and other 'design' features that permit them to continue functioning after absorbing a certain amount of abuse. A dozen rivets, or a dozen species, might never be missed. On the other hand, a thirteenth rivet popped from a wing flap, or the extinction of a key species involved in the cycling of nitrogen, could lead to a serious accident.[4]

One problem with likening species to rivets, however, is that it implies there is a design, that species are deliberately placed where they are needed. Yet our current understanding of how communities are assembled says that species move in and evolve as opportunities present themselves; in other words, when they can, not when they are needed. The role of a species may also change from one community to another or with shifting environmental conditions.

In 1992, Brian Walker of the Commonwealth Scientific and Industrial Research Organization Division of Wildlife and Ecology in Australia recast the analogy and proposed instead that ecologically, most species are more like passengers on the plane than rivets. Only a few species are irreplaceable on a day-to-day basis, and these he called "drivers." His analogy, known as the redundancy hypothesis, suggested that the loss of passenger species will cause little change in the system. That's because most species can theoretically be sorted into functional groups or guilds, a term reminiscent of medieval craftsmen's unions that's used to group ecologically equivalent species, which are species that exploit similar resources, such as nectar or space. The idea is that most functional groups contain enough overlapping skills that if one or a few members are lost, others can take up the slack and get the work done.

Walker's proposal, which has drawn fire from many conservationists, was not meant to serve as a death sentence for the earth's passenger species. However, in a world where funding for conservation is limited and human pressures on natural systems are growing, Walker asserts that a policy placing "equal emphasis on every species is ecologically unsound and tactically unachievable." The goal of conserving biodiversity can be achieved most effectively, he proposes, by focusing on preserving "the integrity of ecosystem function." This means priority should be given to functional groups with little or no redundancy. These groups and species do jobs few others can do, or possess unique talents whose loss would harm the integrity and resilience of the system.[5]

While both the rivet and redundant species hypotheses serve to stimulate discussion and debate, neither provides specific guidelines for gathering information needed in conservation and management decisions. For one thing, scientists don't know who the drivers or keystones or critical rivets are in most natural systems. (That's one reason some conservationists prefer to focus on "umbrella" species instead. These are species such as the spotted owl in the forests of the U.S. Pacific Northwest whose habitat needs are extensive enough that protecting them will likely provide a haven for many other species as well.) Also, no one has looked at specific communities and tried to sort their resident species into functional groups according to their impacts on plant productivity, nitrogen cycling, local rainfall patterns, soil erosion, pollination, pest control, or other services, much less determined how much interchangeable talent any group contains.

(Many ecologists have explicitly discontinued use of the term "redundancy" because it is often used in other management or policy arenas to imply expendability. In ecology, it was intended to have a more positive connotation, much like a "deep bench" in sports, where successful teams maintain strong backup players for each position, or like the backup systems built into reliable technologies.)

As part of the Global Biodiversity Assessment sponsored recently by UNEP, ecologists involved in the SCOPE Program on Ecosystem Functioning of Biodiversity attempted to integrate various ideas about species roles into a general framework, outlining specific kinds of information that may help predict whether the loss or addition of any given population of a species will alter ecosystem processes. Useful prediction, these researchers concluded, must be based on at least five critical types of information:

- How many species does the community contain?
- How many other species in that community are similar in function to the one invading or in danger of local extinction?
- Is the threatened species numerous or rare?
- Does the species that might be lost or added possess unique or influential traits?
- What ties, direct and indirect, does that species have with others in the community?[6]

Each of these types of information can provide ecologists with clues about how changes in biodiversity will affect the dynamics of communities and the workings of ecosystem processes.

How Much Diversity Does an Ecosystem Need?

Ecologists have struggled for decades to explain why a handful of species can create a self-sustaining ecosystem in some regions while communities in other places contain thousands, or tens of thousands of creatures. So far, however, tallying the numbers of species present in healthy natural communities around the earth has provided few indications about how many species are actually needed to generate a full range of ecosystem services.

The most consistent pattern to the earth's biodiversity is that species numbers peak in the tropics and decline toward the temperate regions and the colder, drier ends of the earth. For instance, sandwiched between sandstone crystal in the Ross desert of Antarctica, the coldest, driest, windiest region on the planet, self-contained lichen communities require only half a dozen microbial members to carry out all the processes expected of ecosystems. These communities trap meltwater or capture water vapor from the air; gather nitrogen from the ammonium and nitrate-laden dust that settles onto the rocks; photosynthesize, grow, reproduce, decompose, and recycle nutrients. They have persisted, apparently little changed, for 10 million years or more.[7,8]

In contrast, in the moist tropical forests of both western Amazonia and Borneo, more than three hundred large tree species have been

counted in a single hectare.[9] No one has yet tallied the mass of vines, orchids, and other epiphytes clinging to these trees; the shrubs and herbs
in the understory; the birds, beetles, ants, monkeys, peccaries, jaguars,
and a myriad of other creatures that make their living in the lush and
constant greenness of these communities.

Many theories have been devised to try to explain why biodiversity
reaches such a stunning crescendo in the amenable and changeless climate of the tropics. Among the most convincing is the notion that over
millions of years, species may have become so narrowly specialized that
more of them are able to pack into the neighborhood without displacing
others. But no clear set of answers can yet explain what controls the extent of biodiversity in any given place.[10]

The real question here is: Does the full abundance of species that exists in a community need to be preserved in order to maintain ecosystem processes? A tropical forest produces more than three hundred
times the weight of plant material in the same space as a lichen community on a polar rock, largely thanks to the extra heat, water, sunshine,
and nutrients available.[11] But could the forest be just as extravagantly
productive with ten kinds of trees, or one hundred?

Some ecologists propose that lavish diversity makes little contribution
to the everyday operations of an ecosystem. Beyond some relatively low
threshold number, they say, services such as plant productivity and nutrient cycling gain little from the labors of more species.[12] Recent studies suggest they may be right, at least over the short term.

One such study, an analysis of soil fertility, was conducted on a two-
and-a-half-hectare site in Costa Rica by a team led by Jack Ewel of the
U.S. Forest Service Institute of Pacific Islands Forestry in Hawaii. The researchers first felled and burned the trees, mimicking the larger land
clearing that goes on for agricultural purposes, then divided the site into
a series of small plots. On one set of plots, they allowed natural vegetation to resprout as it would after a blowdown or lightning fire. After five
years, those plots contained more than one hundred species of plants.
Another set of plots was treated much the same way, except that the researchers "enriched" the diversity by scattering thousands of seeds each
month. Five years later, those plots contained two dozen more species
than the plots allowed to regrow naturally. On a third series of plots, the
researchers planted monocultures of different crops each year: maize,
cassava shrubs, then, finally, a fast-growing native tree.

As the team expected, the two types of high-diversity plots retained soil nutrients much better than the single-crop plots, although losses in soil fertility declined to very modest levels once a tree crop was planted. Yet somewhere between one species and one hundred, the benefits of plant diversity for soil fertility leveled off. The extra two dozen species in the enriched plots had no effect on nutrient cycling.[13]

As part of the SCOPE project, Joseph Wright of the Smithsonian Tropical Research Institute in Panama reviewed Ewel's evidence and also compared published results on productivity in high-diversity forests with those obtained in various tropical timber plantations and low-diversity forests. Altogether, he found little evidence that plant productivity requires the full abundance of plant species found in the tropics. He reasoned that, although many tropical forest plants fill very narrow ecological niches, others possess broad talents. Thus, in a place as species-rich as the average tropical rain forest, a high degree of ecological redundancy is inevitable. Because of this, Wright predicts that plant productivity tops out "at levels of plant species richness well below those commonly observed in tropical forests."[14]

Exactly what the threshold number of species required for full performance is, for each function in each ecosystem, remains unknown. Ecologists suspect that the identity and talents of the species present matter more than their numbers per se. In other words, the key lies in having a complete array of species traits—enough plant architectures and life strategies, for instance, to capture every bit of sunlight, sop up every nutrient, exploit moisture from every layer of the soil, stabilize the soil against erosion, and so on. As soon as all tasks are covered, either by many species with narrow specialties or fewer species more broadly adapted to a wide array of environmental conditions, then adding more species will have less and less impact. In fact, Stanford ecologists Peter Vitousek and David Hooper predict that the biggest effects on nutrient cycling and perhaps productivity will come with the first one to ten plant species in the system.[15]

One conclusion to be drawn from these analyses is that species losses or additions may cause the biggest impacts in communities where numbers are already low, areas such as polar deserts, the Arctic tundra, or oceanic islands. Even so, the high species numbers in lush communities cannot be viewed as unnecessary frills. Rather they may provide critical insurance over the long term, supplying backup talent and a wide range

of tolerances that help to average the ups and downs as ecosystems resist and rebound from fires, hurricanes, floods, droughts, pest outbreaks, and climate shifts.[16]

The relationship between numbers of species and stability has, in fact,
been one of the longest-running debates in ecology. In the 1950s, pioneering ecologists Charles Elton of Oxford and Robert MacArthur of
Princeton each proposed that species-rich communities are more stable
than species-poor communities. The concept, widely accepted at first,
became controversial, in part because stability meant different things to
different scientists. The term was used to refer to the phenomena of resistance to invasion by new species, population ups and downs, changes
in a community's species mix, or resilience in the face of shocks and
disturbances.

In the early 1970s, mathematical analyses by Robert May, now chief
science advisor to the U.K. government, indicated that more diverse and
complex assemblages might actually be less stable.[17] May also hinted at
the existence of specific cases where diversity might enhance stability.
He wrote as an "afterthought" to his 1973 analysis that "if we concentrate on any one particular species our impression will be one of flux
and hazard, but if we concentrate on total community properties (such
as biomass in a given trophic level) our impression will be of pattern and
steadiness." In other words, species diversity might support just the kind
of stability with which this book is concerned—the stability of ecological services over time.

One of the few experiments to investigate this possibility was reported
in 1996 by David Tilman of the University of Minnesota, St. Paul. Results
of his eleven-year study in the prairie grasslands of the United States
Midwest indicate that diversity stabilizes an ecosystem and its
processes, but does so at the cost of the individual species that live in
these communities.

Tilman's study cataloged the year-by-year production of plant material in 207 grassy plots at the Cedar Creek Natural History Area in Minnesota. Each plot contained anywhere from a handful to two dozen
species of prairie grasses and wildflowers. Through all the fluctuating
environmental conditions, which included a severe drought as well as
the usual variations in wet and dry years, biomass production in the
most diverse plots was much steadier than in the species-poor plots. Yet
in the more diverse plots, the populations of each species swung widely
from one year to the next, much more so than those in the species-poor

patches. Diversity thus offers little succor to individual species, subjecting them instead to greater competition. But when the work of one species is hampered by drought or other conditions, another species in a diverse and multi-talented community is likely to thrive and expand its activities. It is this ability of species to compensate for one another that allows the system as a whole to function steadily over the long term.[18]

Ecological Backup Players

If many species in an ecosystem perform the same function, theoretically the community should be better able to compensate for the elimination of one or the addition of another without a change in operations. Few experiments have been done to test this idea, partly because it crosses the lines of academic disciplines. Researchers must not only monitor ecosystem processes over time, but must also document changes in populations and the species mix in a community. Yet, in several experiments in aquatic systems, just as in the Minnesota grasslands, ecologists have watched compensation take place, as populations of some species declined while others increased.

In one instance, ecologists Bruce Menge and Jane Lubchenco of Oregon State University and their team set up experiments along the barren, rocky shores of the small tropical island Taboguilla in the Bay of Panama. Their mission was to learn whether there was a consumer—grazer or predator—in the intertidal zone whose loss would alter life for everyone in the community. Their plan was to eliminate various groups of animals one at a time and monitor the results. First, the researchers sorted the animals of the community into four functional groups. One slow-moving group was dominated by predatory whelks, a second by grazing limpets. Among the fast-moving consumers, a third group included mostly crabs and small fish such as blennies, and the fourth assemblage was dominated by large omnivorous fishes—damsel fish, chub, parrot fish, wrasse, and porcupine fish.

The team found that deleting all of these consumers by caging off the rocky shore where they feed changes the rock community drastically. Boulders once covered almost entirely by flat crusts of algae acquired a lush lawn of green algae, then red seaweeds moved in over time, and finally barnacles and rock oysters. Excluding only one of the four groups of consumers exerted almost no effect. When Lubchenco removed just

limpets, for instance, a lush and diverse crop of algae grew up. Yet grazing fish quickly scraped the algae off the rock, taking over the work of
the grazing limpets. The team's results indicated there was no keystone
consumer along that shore, no organism whose role could not be filled
by others. Delete one group and others stepped in to claim the freed resources and mask any systemic effects of the loss.[19]

Interesting results have also been obtained from fresh water lakes. For
example, limnologist Thomas Frost of the University of Wisconsin and
his colleagues found that acidification caused major declines in some
species of zooplankton but had no apparent effect on total biomass of
these creatures. Over a seven-year period, the researchers partitioned
and slowly acidified one side of Little Rock Lake in Wisconsin to examine the impact of acid rain on zooplankton, the mostly microscopic animals that serve as the major food source for trout and other game fish.
As pH levels in the lake dropped, the team began to see major declines
in some types of tiny rotifers, water fleas, and crustaceans called cope-
pods, which prey on small grazers. Despite these losses, however, the
total weight of zooplankton in the lake remained fairly stable (at least
until acid levels hit 4.7) because other species, including some rare types
in each of these animal groups, increased. The result was a very different zooplankton community in the acidified side of the lake, yet one that
seemed to maintain the same role in the food web.[20]

In spite of such results, scientists have found it difficult to generalize
from one community to another. In some rocky intertidal communities,
keystone predators have been identified, and other consumers cannot
compensate for their loss. Other lakes with a different water chemistry
and simpler food chain seem to have much less capacity to make up for
species losses than Little Rock Lake. For instance, experimental acidification of a small lake in Ontario, Canada, turned up much less functional redundancy. Half of the lake's original species were lost, and only
a third were replaced. Among the acid-sensitive species whose place in
the food chain remained unfilled were a small fish and two tiny crustaceans that were major food sources for lake trout, the top predator in
the system. When these three species were lost, trout populations collapsed. David Schindler of the University of Alberta, Canada, suggests
that species-poor lakes in northern alpine regions might have limited capacity to adapt to changing conditions. This could explain, for example,
why Norwegian lakes have been so dramatically affected by acid rain.[21]

Even in the Little Rock Lake study, Frost and his colleagues found they couldn't have predicted exactly which rare species would fill in for the lost zooplankton species.[22] Indeed, scientists may never be able to predict in advance which species are critical backup players that ought to be protected. Findings like these should certainly cause conservationists and decision makers to think twice about allowing rare species to go extinct. The fact is that their talents may not become apparent until both the community and the environment have changed.

Moreover, even a strong backup team of extra species cannot carry an ecosystem through every natural disaster or human act of abuse unscathed. Steve Carpenter of the University of Wisconsin points out that, while moderate stress usually changes only species composition, more severe stress eventually alters ecosystem functioning. This pattern appears to hold for many different stresses in a variety of lakes.[23] When the pH of Little Rock Lake dropped below 4.7, even the backup species faltered, and real damage to ecosystem processes began to occur. "The lesson here is that species losses at moderate levels of stress should serve as a miner's canary," Carpenter says. "They indicate that there are likely to be consequences at the ecosystem level if the stress is intensified."

Compensation has other limits, too. The fact is that species fill multiple roles and no two species can be considered fully ecologically interchangeable, especially over the long term. If kelp beds are destroyed, planktonic algae may compensate by keeping up primary production, but these tiny plants cannot re-create the kelp forest habitat that provides a home to fish and marine mammals and protects the coastline from erosion.[24]

When an introduced blight eliminated the chestnut trees that once dominated southeastern U.S. forests, other species such as yellow poplars were able to fill out the canopy again, although it took nearly half a century. No cascade of extinctions followed, and it may appear that other forest species have fully compensated. But researchers have found some evidence that the poplar-dominated forest may not be as resilient as the one it replaced.[25]

Like many effects of biodiversity loss, the full impact of the loss of chestnuts may take several human lifetimes to manifest. Unfortunately, land use and conservation policies in most parts of the world are too shortsighted to deal with such effects. The result is that one generation often unwittingly sets in motion a deterioration of ecological services from which the succeeding generations must inevitably suffer.

Identifying Keystone Species

The concept that a single species might hold the key to both the diversity and the stability of its community was first outlined in 1966 by ecologist Robert Paine of the University of Washington. He was seeking to explain the pattern of diversity he observed among species clinging to the rocks in the intertidal community along the Pacific Coast of Washington. The community Paine saw along these temperate shores teemed with mussels and goose-necked barnacles high on the tidal-washed rocks and with anemones, chitons, limpets, sponges, nudibranchs, and various algae lower down. The top predator was the seastar *Pisaster ochraceus,* which devoured a wide array of invertebrates ranging from mussels to snails, barnacles, chitons, and limpets. Its primary targets, though, were mussels.

In fact, Paine found the seastar preyed on the mussel so skillfully that it effectively kept this highly competitive species from monopolizing space on the rocks. When Paine removed all seastars from some sections of the shoreline, the result was a dramatic drop in diversity because the mussels, their expansion unchecked, took over the rocks and crowded out other organisms. The other predators in the community, such as snails, crabs, and other species of seastar, simply weren't voracious enough to compensate for the loss of *Pisaster* and curb the mussels' "monopolistic tendencies," as Paine called them. As a consequence, the number of species in the community fell by nearly half, from fifteen to eight.[26,27]

Perhaps the best-known keystone predator to be identified so far is the sea otter, a whiskered, antic, fascinating mammal that forages in the nearshore zone. Researchers didn't have to exclude otters from certain regions to test their influence over the community, since hunters had effectively done so. During the eighteenth and nineteenth centuries hunters virtually eliminated the otter throughout most of its historic range, which ran from the Aleutian Islands off Alaska to Baja, California.

James Estes of the University of California, Santa Cruz, and John Palmisano, then of the University of Washington, set forth to compare two areas in the Aleutian Islands. Around one island group, otters have remained largely absent since the eighteenth century. At another island group four hundred kilometers away, otters were thriving along the coastline, as many as twenty to thirty animals per square kilometer of habitat. (By the early 1970s, with legal protection, otters had begun to recolonize parts of the coastline and were being reintroduced in other places, despite opposition from commercial abalone fishermen, who view otters as strong competitors.) Estimated conservatively, that number of otters would be consuming thirty-five thousand kilograms of fish, sea urchins, and abalone per square kilometer every year.

Much of the study revolved around the status of one of these prey species, the spiny sea urchin, *Strongylocentrotus* sp. Where populations of these mobile pin cushions are dense, they devour vast areas of kelp forest and seagrass beds, creating aquatic moonscapes known as "urchin barrens." The researchers realized that in areas where urchins had been experimentally removed or killed by oil spills, kelp forests regrew rapidly. Had sea otters once been keystone predators that had held the urchins in check and allowed kelp to flourish?

Sure enough, around the otter-patrolled islands, the researchers found the urchins were small and inconspicuous. As a consequence, the kelp forest was luxuriant. Yet, around the islands without otters, kelp was distinctly absent, and urchins formed a literal carpet across some parts of the sea floor, relieved only by extensive beds of mussels and barnacles. Because of these results, the researchers proclaimed the otter a keystone predator, powerful enough to control the structure of its nearshore community.

Since a thriving kelp forest, anchored on the bottom and stretching to the surface, supports a food chain that affects many other creatures, Estes and Palmisano also speculated that it was no coincidence that islands without otters also lacked harbor seals, bald eagles, and game fish known as rock greenling.[28]

Others have since found that otters exert a strong influence, not only in rocky habitats but also in soft-bottom coastal areas where they range. Here they make their mark not just through predation, but also through their role as disturbers, digging foraging pits in the sandy bottom in search of butter clams and other bivalves. In the process, they stir up bottom sediments and litter the surface with piles of exposed and discarded shells where kelp, anemones, and other organisms can find firm attachment sites. Thus, they change the species mix in these communities, too.[29]

Moving from the water to the land, one question about keystone predators seems obvious: Are any of the large, fierce carnivores of the African savannas—the lions, cheetahs, leopards, wild dogs, and others— keystones? Does the presence of one or more of these predators reduce competition among antelope and other prey and allow more species to coexist? Unfortunately, the question has never been formally tested. Anecdotally, however, scientists do know that in some game parks in southern Africa where large carnivores have been eliminated, park rangers have had to take over the role of predator, culling abundant species such as impala to reduce the competitive pressures on rare species such as roan antelope.[30]

The most thoroughly studied keystone predators on land are much less charismatic than lions and cheetahs, and they consume seeds rather than flesh. They are nocturnal desert rodents known as kangaroo rats. James Brown of the University of New Mexico and his colleagues fenced off quarter-hectare plots in the Chihuahuan Desert of southeastern Arizona with fine wire mesh and removed all three species of resident kangaroo rats. After twelve years without kangaroo rats, the habitat changed dramatically, from shrub-dominated desert to grasslands. Despite the continued presence of harvester ants, mice, and other seed-eaters, the removal of the three kangaroo rat species, which Brown labeled a "keystone guild," allowed tall, large-seeded grasses to expand considerably in the open ground between the shrubs. Like sea otters that disturb bottom sediments, kangaroo rats disturb enough soil when they forage, burrow, and cache seeds to discourage the colonization of tall, long-lived grasses. It was the elimination of both their seed-eating and their soil-churning ways that allowed the grasses to flourish.[31]

Organisms that physically shape their community are known as "ecosystem engineers," a term coined by an international team of ecologists—Clive Jones of the Institute of Ecosystem Studies in southeastern New York, John Lawton of Imperial College in the United Kingdom, and Moshe Shachak of Ben-Gurion University of the Negev in Israel. Some of these engineers, like kangaroo rats and sea otters, change their environment through their physical activities, such as tunneling, pecking holes, bulldozing or gnawing down trees, stomping trails, gouging out pits and wallows, or rototilling the soil. Others, such as reef-building corals, modify or literally create habitat through their own physical structure. Whether they engineer by doing or being, a number of these organisms have been nominated by various researchers as keystones.[32,33]

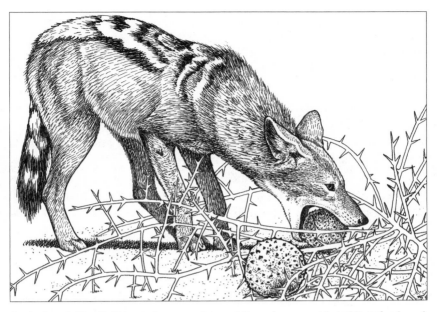

In the harsh Namib Desert, the nara vine and its melons provide habitat, food, and water for creatures from jackels to dune larks.

One of the organisms that changes the community just by existing is the nara vine. On the nearly barren dunes of the hyper-arid Namib Desert along the southwest coast of Africa, the massive roots of the nara vine anchor an entire "micro-ecosystem." Individual plants may live a century or more, clinging to the margins of ephemeral rivers and sending out fifty-meter-long roots to tap subsurface water deposits. Like coral reefs in the sea, these spiny, leafless plants provide habitat as well as food and water for animals ranging from insects to ostriches to humans. Nara shares a place in the cucurbit family with cucumbers and gourds, and it produces a melon that's been called a "desert canteen." Not only are its melons 80 percent water by weight, but for more than eight thousand years, humans have harvested the fruits for their carbohydrate-rich pulp and nitrogen- and phosphorus-packed seeds. Gerbils nest under the vine and feed on the beetles and windborne plant litter it collects. Dune larks nest in its thickets. Ostriches, springbok, oryx, and lizards nibble its stem tips. Jackals, hyenas, and porcupines feast on the melons. Thus, in a strictly passive fashion, this vine creates a unique ecological refuge for many species.[34]

A larger number of organisms have been nominated as keystones because of the physical disturbance they create in their environment. African elephants, for instance, are able to convert vast areas of wooded savanna to open grassland by toppling and ripping apart trees and shrubs as they browse. By virtue of their foraging behavior, elephants maintain open grasslands in a way that bulldozers, plows, drag-chains, and other human engineering tactics have had little success in mimicking as range managers try to reverse the encroachment of woodlands and shrub deserts into once-productive grasslands.

Porcupines have also been labeled keystones in various woodlands in southern Africa because they prefer to feed on certain dominant trees, stripping away the outer bark to feast on the inner layers. These damaged trees are more likely to be killed by fires or disease, and thus the ravages of porcupines help maintain a diverse and patchy landscape of grassland and thickets.

Hippopotamuses and buffalo in Africa and alligators in North America create mud wallows that turn into marshy pools, diversifying the habitat for many other creatures.[35] Beavers fell trees and create dams, ponds, and extensive wetlands that alter the flow of water, the transport of sediment and nutrients, and the diversity of plant and animal life—effects that can alter the landscape for centuries.[36]

Animal engineers don't have to be large to have major impacts on an ecosystem. Consider three related species of herbivorous snails in the central highlands of Israel's Negev Desert. Some 70 percent of that desert surface is covered with limestone rock, and only 30 percent with soil. Lichens live within the crystals of these rocks, just as they do in the Ross Desert of Antarctica, and they, too, take up nitrogen in the form of dust mixed with dew that has settled on the rock. The snails make their living rasping up and devouring both the rock surface and the lichen within, then excreting the rock grains. Their feeding weathers away the rock at the rate of more than a ton per hectare per year, and their feces create as much new soil as is deposited by windborne dust, which was previously considered to be the major source of new soil in the desert. The feces, deposited on the soil beneath the rocks where the snails shelter during the heat of the day, also supply a significant proportion of the nitrogen that fertilizes the growth of higher plants in these desert soils. Presumably, the desert would be much less green without the work of the snails.[37,38]

Elsewhere, pocket gophers, badgers, aardvarks, rabbits, crabs, moundbuilder or leaf-cutter ants, and termites perform similar functions, concentrating or redistributing important resources around their mounds and tunnels.[39] Small burrowing crabs that populate coastal mangrove forests in many tropical regions, for instance, are often voracious consumers of mangrove seeds and leaf litter, greatly speeding up decay and nutrient turnover. In experiments where crabs were removed from sections of red mangrove forests off north Queensland, Australia, the loss of their burrowing and soil-aerating activities caused potentially toxic sulfide and ammonium levels to rise in the soils and slowed tree growth and reproduction.[40]

What of other interactions between species that might involve keystones? Pathogens and parasites, despite their tiny size, can shape the community and the landscape as thoroughly as seastars, otters, and elephants. The tsetse fly and the sleeping sickness organism that it vectors have been enormously effective in keeping humans from settling vast areas of Africa. The flies also seem to affect the behavior of other large mammals besides humans. In one game reserve in Zimbabwe, for instance, elephants had avoided certain wooded thickets, apparently because of the heavy infestation of tsetse flies. When officials eliminated the flies with insecticidal sprays, the elephants moved in and within a few years had converted the woodland to open savanna.

Beneficial interactions such as nitrogen-fixing services, pollination, or seed dispersal—relationships that fall into the category of mutualisms—may also generate keystones. Some nitrogen-fixing bacteria that live in nodules on the roots of plants play crucial roles in the structure and functioning of communities in nutrient-poor settings. So might mycorrhizal fungi that associate with plant roots and supply phosphorus and other nutrients. Some tropical trees such as figs and cecropias that produce fruits that nourish animals in seasons of scarcity have also been nominated as keystones.[41]

As this litany of proposed keystones indicates, twenty-five years after Paine coined the term, it was being applied widely. Too widely, some ecologists believed. They were concerned that fuzzy definitions might cause the keystone concept to be misused in conservation decisions.[42,43] But others considered the term "too charismatic, too entrenched, and too useful to be abandoned."[44] In late 1994, as part of the Global Biodiversity Assessment project, SCOPE called together more than a dozen ecologists in Hilo, Hawaii, to reconsider what is known about the forces that organize communities. The group proposed the following definition of a keystone species: "One whose impact on its community or ecosystem is large, and disproportionately large relative to its abundance." That distinguishes keystones from dominant or abundant species, such as redwood trees and coral reefs, which have large impacts but not out of proportion to their commanding physical presence in their communities.[45]

What generalizations can be made about keystones? The Hilo group decided that only a small proportion of the organisms in a community are likely to be keystones. Such singular species are likely to be found in every major type of ecosystem. Although many may be at the top of the food chain, keystones may make their mark on the community through almost any type of interaction, from competition to pollination, disturbance, or disease.

Still, ecologists have a long way to go before they can deduce from general principles which species in any given community are keystones. Up till now, most keystones have been identified through the painstaking observations of ecologists well versed in the natural history of the system. Some cases, like that of the sea otter, relied on natural experiments created unwittingly by overhunting or other human exploitation. Deliberate manipulations, such as the exclusion of seastars and other intertidal consumers, provide much useful information. Yet such experiments are neither practical nor ethical in all cases. Moreover, such studies may be too time consuming. Trying to remove a single pollinator and then watching for system-level repercussions in a tropical forest, for instance, would, in addition to the logistical problems, require decades or more for any effects to become clear. Such a timetable becomes unworkable when scientists are faced with pressing management or conservation decisions.

Another problem with venturing into the field and picking out a keystone species is that if the system is working right, even a top predator and its potentially powerful prey might both be quite rare. Paul Ehrlich cites the example of heading into the field in Queensland, Australia, in the early 1970s and looking unsuccessfully for caterpillars of a tiny moth called *Cactoblastis cactorium* that feeds on the scattered clumps of imported prickly pear cactus. Without knowing a bit of history, he might have assumed that neither the cactus nor the elusive moth was an important player in that ecosystem. Yet the cactus, introduced to Australia from South America in the mid-nineteenth century, had invaded Queensland so thoroughly by the turn of the century that it nearly destroyed cattle ranching in the region. Relief came when entomologists went searching in South America for a natural enemy of the cactus and returned to Australia with *Cactoblastis*. The moths have since consumed enough cactus to render both the plant and themselves rare. Yet if that moth were to disappear, Queensland could find itself blanketed in thorn once again.[46]

Sometimes human activities have so skewed the natural order in a community that spotting the keystone even requires excluding humans from the system. Along the rocky shoreline of Chile, divers and gatherers each year harvest tens of thousands of tons of a carnivorous whelk they call "loco" (*Concholepas concholepas,* known as "Chilean abalone") for the seafood market. Not until researchers Juan Carlos Castilla and L. R. Duran of Pontificia Universidad Catolica de Chile located a relatively unperturbed section of the coast and excluded human predators for two years did they uncover the whelk's role as a keystone predator. The restoration of high numbers of whelks not only led to a dramatic decline in mussels but an overall increase in diversity as other creatures reclaimed space the mussels had once monopolized.[47]

There is another, more complicated twist to the whelk story. *Concholepas,* it turns out, has keystone status only where mussels and barnacles dominate the community. That's because, to control the allocation of space in the community, the whelk must be able to bulldoze its prey off the rocks, as well as devour it. Along stretches of the coast where sea squirts dominate the rocks, whelks cannot bulldoze them and so cannot fill the keystone role.[48]

Increasing evidence, in fact, shows that the effects of whelks and other keystones are what ecologists call "context dependent." In other words, a creature's ecological importance—the strength of its interactions with others—varies in space and time. This reinforces the idea that function is an emergent property of a creature interacting within a system; it is not a set of fixed talents sculpted by natural selection and expressed by an organism at all times, places, circumstances. Rather it depends on factors such as the level of disturbance in a particular habitat, the mix of other species in that community, and even historical events.[49]

Even *Pisaster,* the original keystone, is just another seastar along some stretches of rocky coastline. Recently a research team led by Menge of Oregon State reexamined the interactions between *Pisaster* and its primary prey, mussels, at several sites along the Washington and Oregon coasts. On rocky headlands exposed to pounding waves, the seastars consistently maintained their keystone roles. But their influence diminished just tens of meters away in wave-sheltered sites, where mussel populations grew more slowly and where whelks and possibly other predators also sought out mussels. In coastal areas where sand regularly washed over the rocks, sand burial was apparently the primary cause of mortality among mussels, not seastars.[50]

Factors as tenuous as history may also play a role in making or breaking keystones. Take the case of two islands, Malgas and Marcus, along the west coast of South Africa. Lobsters abound in the waters around Malgas Island, where they readily consume mussels and most whelks, serving as keystone predators and preventing mussels from dominating the community. The lobsters, however, avoid two species of whelks. Why such picky tastes? South African researchers Amos Barkai of the University of Cape Town and Christopher McQuaid of Rhodes University found that the whelks are unpalatable to lobsters because they acquire a thin crust of tiny colonial animals called bryozoans on their shells.

Until the 1960s, Marcus Island looked much like Malgas. No one knows what caused lobsters to disappear from Marcus. What Barkai and McQuaid wondered was: Why haven't lobsters recolonized? They got an unexpected and rather ghoulish answer when they took one thousand lobsters from Malgas and turned them loose at Marcus. The whelks immediately turned the tables on the returning, would-be keystones, attaching themselves by the hundreds to each lobster, much as army ants attack a lizard, weighing them down and devouring them. Within a week, not a single live lobster could be found. Once freed from the lobsters' dominance it seems, the whelks had burgeoned to invincible numbers that prevented a return to the old order.[51]

Supporting the Community Web

One of the most important but least predictable consequences of losing species may be the feedbacks or reverberations that spur further changes in community structure and ecological functioning. By definition, the loss of a keystone species sends a dramatic ripple throughout a community. Even species that are rare or seemingly inconsequential may exert large indirect impacts on ecosystem processes through their interactions with other species.

Peter Raven of the Missouri Botanical Garden has estimated, for instance, that every time a plant goes extinct, an average of ten to thirty other species also collapses, like a house of cards.[52] That's especially true in the tropics, where the lush and uniform greenness of the vegetation cloaks a diverse pharmacopoeia of defensive chemicals in leaves, seeds, stems, and other plant tissues. Many herbivorous insects have evolved highly specialized feeding strategies that allow them to overcome the toxins of specific plants. Each insect in turn supports tiny parasites. In this way the tropical forest has been subdivided into a multitude of semi-independent food webs.

However, there are generalized services that many tropical plants require, such as pollination and seed dispersal. These are often provided by what Lawrence Gilbert of the University of Texas at Austin has called "mobile links": Animals such as bees, moths, bats, and hummingbirds that perform functions critical to the reproductive success of a wide array of plants, each of which may support a largely independent food web.

Among the most important groups of mobile links in the tropics are the brilliantly colored euglossine or orchid bees, long-distance fliers that are the only pollinators of certain rare or widely scattered plants. The males of each species pollinate orchids while gathering from them specific volatile chemicals, which they use as perfumes to attract females. Orchid bees of both sexes also visit hundreds of plant species at all successional stages, from the understory to the forest canopy, in their search for pollen and nectar as well as resin with which to build nests. So the survival of these bees—and in the long run the orchids that rely on them—depends on an extensive mosaic of natural tropical forest habitat.[53]

Once again, this example reinforces the notion that large numbers of species do not necessarily signify a high degree of functional redundancy. Although tropical rain forests can lose a significant proportion of their plant species and still maintain their productivity—that is, the sheer tonnage of greenery—plants are not fully interchangeable to the animals that rely on them for pollen, nectar, resin, volatile chemicals, fruits, seeds, foliage, and other properly timed and palatable resources.

Despite the gaps in our knowledge of how species influence communities and larger ecological processes, some general patterns seem clear:

- The revised view of keystones suggests there may be more indispensable drivers or critical rivets than ecologists once imagined, even though researchers have no tidy checklist for picking them out.

- Scientists can be fairly sure that eliminating a dominant or abundant species will change the character of a community and probably the functioning of an ecosystem.

- Most systems contain some redundancy, some functional overlap between species, but no means exists for predicting which of the minor players in a community may step in to compensate for lost ones.

- Communities with relatively few species are likely to be strongly affected by losses or additions.

- Even in species-rich communities, biological losses may reverberate, leading to further impoverishment and an eventual compromise in ecological services.

The following chapters provide many examples where species losses or additions have led to changes in the structure of the community and the integrity of vital ecosystem processes.

CHAPTER THREE

Community Ties

Overleaf
Without jaguars to prey on peccaries and other seed-eating mammals, the mix of trees and shrubs in tropical forests would shift.

$\mathscr{A}$ unique autumn spectacle along McDonald Creek at one time drew up to forty-six thousand visitors to Glacier National Park on the U.S.–Canadian border. In the early 1980s, tourists who flocked to viewing areas along the four-kilometer stretch of creek could expect to see hundreds of bald eagles congregated on stumps and trees while grizzly bears stalked in the shallows. The eagles and bears were drawn to the creek by kokanee salmon, as many as one hundred thousand that annually swam upstream from Montana's Flathead Lake to spawn. Gulls, mergansers, mallards, goldeneyes, and dippers, along with coyotes, minks, and river otters gathered, too, feasting on the kokanee or their eggs.

By 1989, however, the show had ended. Only fifty kokanee arrived to spawn that year, and half that many eagles stopped briefly on their journey to wintering grounds further south. Human anglers caught not a single kokanee.

When researchers began investigating the collapse of the kokanee population, they found that it was one of a series of changes that had cascaded through the Flathead Lake ecosystem within twenty years after state fisheries officials began stocking exotic opossum shrimp into upstream portions of the watershed. Ironically, small shrimp were intended as extra food to help boost kokanee numbers. The plan overlooked an important bit of the natural history of both shrimp and fish: Kokanee feed on zooplankton near the surface during the day. Opossum shrimp spend the day near the bottom, largely out of reach of the fish. At night the shrimp migrate upwards to feed on zooplankton themselves—the same zooplankton, unfortunately, that serve as the chief food for kokanee. Rather than supplying a new food resource for the kokanee, humans had unwittingly introduced a competitor.

After opossum shrimp drifted in from tributaries and appeared in Flathead Lake in 1981, zooplankton quickly declined, especially populations of daphnia, or water fleas, which are a favored food of both the kokanee and the shrimp. Within just a few years, the kokanee population in the lake had collapsed too. One hundred kilometers upstream in McDonald Creek, the disappearance of spawning kokanee eliminated a food resource that had once fortified eagles for their winter migration and fattened bears for hibernation. It also brought to an abrupt end a wildlife spectacle that had boosted off-season tourism revenues for the park and neighboring communities.[1]

The immediate effects of species additions or subtractions are often merely the opening acts in a complex drama. As the chain of events in Glacier National Park and the sagas of keystones such as sea otters illustrate, a change in species can not only disrupt relationships among the organisms in a community, but can also set in motion a cascade of secondary or indirect effects that can reverberate through a system for centuries. Eventually this widening circle of disturbance may affect the structure or stability of a community and interfere with the operation of larger ecosystem processes.

It would be misleading to claim that every species loss—or every invasion by a non-native creature—threatens systemwide disruptions. Species can often come and go without affecting large-scale processes, such as nutrient release, soil fertility, plant productivity, pest and disease control, or water cycling. From a human perspective, of course, the precise mix of species may matter a great deal. In places like Glacier National Park and in wildlands around the world, society's goal is to perpetuate diverse and self-renewing natural communities not only for maintenance of vital ecological processes, but also for the genetic or chemical resources they contain, the foods and other commodities they provide, the cultural or spiritual heritage they represent, or the recreational and tourist industries they support.

Tampering with the species mix in a community without causing systemwide ripples might be possible if ecologists had clear formulas for spotting important players and tracking their links with other plants and animals. In the obvious absence of such knowledge, however, humans have received unwanted surprises time and again by over-exploiting valuable species, killing off pesty ones, allowing others to go extinct, and setting loose invasions of native communities by exotic organisms. Sometimes the destruction of species has been so wanton and thorough that the loss of ecosystem services has become apparent almost immediately.

A classic misadventure occurred in the 1950s when Chinese officials grew concerned that flocks of birds were allegedly devouring large amounts of grain. To stop the assault, they declared the sparrow, by which they apparently meant almost any small perching bird, one of the country's major scourges, in company with rats, mosquitoes, and house-flies. With the regimented enthusiasm of Mao Zedong's followers, millions of Chinese went to work killing birds. In one three-day war in Beijing in 1958, near-hysterical crowds are reported to have killed eight hundred thousand birds.

The consequence, of course, was major outbreaks of insect pests. The misjudgment was acknowledged, the sparrow was officially removed from the list of scourges and the killing of insect-eating birds was subsequently banned.[2,3]

Often, human-driven species changes are more subtle than the massacre of sparrows and even quite unintended. Yet the long-term indirect effects can be equally drastic. Predatory sea lampreys from the Atlantic Ocean apparently took more than a century to invade North America's upper Great Lakes after the completion of the Erie and Welland Canals in the 1820s. Once these eel-like bloodsuckers arrived in Lake Michigan and Lake Superior, however, they quickly attacked large, commercially important species, such as lake trout. From that point, it took the lampreys only another twenty years to bring a productive fishery to the brink of collapse.[4]

The importance of indirect effects on community dynamics and the functioning of ecosystems is not well studied. Yet this chapter will point out many cases like those of the sparrows and sea lampreys where human activities have disrupted linkages between organisms, setting in motion domino effects that undermine ecosystem processes. The threatened linkages include all the textbook interactions that ecologists study: mutualisms, such as pollination and seed dispersal, competition, herbivory, predation, parasitism, and pathogenicity. The activities that disrupt these linkages often involve land clearing and habitat destruction or international trade and the introduction of exotic species, including disease agents and pests.

Plants and Their Pollinators

The breakdown of the complex web of interactions between plants and animals is one of the most disruptive but largely invisible consequences of widespread forest clearing and the fragmenting of continuous natural habitats by roads, crop fields, or housing developments. It is what Daniel Janzen of the University of Pennsylvania has called "a much more insidious kind of extinction: the extinction of ecological interactions."[5]

Species that survive initial land clearing find themselves in habitats much reduced in size, surrounded by dissimilar, usually less diverse, and sometimes hostile, neighboring landscapes. Plants that end up at the edge of a formerly continuous forest, for instance, may have to endure unaccustomed wind, sunlight, and desiccation. They may also be exposed to interactions with new organisms, from livestock to exotic pests, and at the same time, be deprived of their traditional associations with insects and other animals.

Indeed, the first casualties may be plants that depend on specific animals for pollination and seed dispersal for their reproductive success and ultimate survival. Loss of these plants—some of them keystones—removes sources of nectar, fruit, seeds, and foliage needed to sustain an array of other animals, from bees to butterflies, bats, birds, monkeys, agoutis, and peccaries. As animals like these grow rare or disappear, their loss threatens the success of other plants that rely on their services. Thus the loss of a single species may result in a widening circle of extinctions and the collapse of ecosystem processes.

The fraying of a complex community web may be quite hard to spot until the damage is well underway. Subtle signs, such as diminishing populations of individual plants or insects, may be masked by largescale depredations like forest clearing and burning. Even in the surviving forests, it could take decades before the lack of saplings of long-lived trees becomes obvious. Researchers willing to explore this potential threat at the tedious level of pollen grains, seed counts, and pollinators' visits to flowers have already found telltale signs of reproductive declines among plants in many regions, from tropical forests to European meadows and North American deserts.

In the dry subtropical forests of the Argentine Chaco, for instance, where large areas have been cleared for crop fields and pastures, Marcelo Aizen of the University of Massachusetts and Peter Feinsinger of the University of Florida compared plants in continuous expanses of forest with the same species growing in forest fragments of various sizes. The researchers found that pollination levels and fruit and seed output declined an average of 20 percent when a species was growing in fragmented habitat. The level of decline was not related to life form or to whether a species could self-pollinate. Whether the plants were herbs, shrubs, cacti, trees, or epiphytes such as bromeliads, the results were the same.

At first this seems odd, because plant species that can self-pollinate should not be affected if numbers of native bees, butterflies, hawk moths, hummingbirds, and other pollinators decline. Obviously, the loss or replacement of pollinators is not the full story. The very fact that even self-pollinators declined indicates that there are multiple stress factors at work in these fragmented habitats.[6]

In North America's Sonoran Desert, stately agaves, also known as century plants, take decades to mature to flowering stage. Mature plants send up a single three- to seven-meter flower stalk that may support as many as thirteen hundred buds. The buds open in the evenings, offering nectar and exposing their pollen to flocks of nocturnal, nectar-feeding bats. After a single flowering, agaves die rapidly, leaving their seeds to dry in pods and be shaken out by wind or visiting birds. If pollination was successful, those pods may contain three-quarters of a million seeds. When bats become rare, that number can drop by 95 percent.

Seed production by agaves in the Sonoran Desert has been dropping, paralleling a decline among their bat pollinators.

Even under the best circumstances in this harsh land, only one or two out of three-quarters of a million seeds may germinate and survive. This means that the consequences of impoverished seed production season after season may take decades or more to show up as the naturally sparse crop of young agaves grows ever sparser.

Sadly, seed production by the agaves has been dropping for thirty years, paralleling a decline in their bat pollinators. Which came first, the decline in bats or agaves, is hard to say, but now each helps drive the other downward. The loss of bats is apparently driven by habitat destruction, especially in southern Arizona, but perhaps also by human exploitation of the agaves that nourish them. Since pre-Columbian times, these spiky plants have been harvested by people for food, fiber, and alcoholic beverages, a practice that continues today in northern Mexico.

The agaves provide food, water, and shelter for a variety of desert creatures, so their eventual loss would have broad repercussions for the desert community. Diminishing populations of agave could cause further declines of bats, which at other seasons are key pollinators and seed dispersers for giant saguaro and organ-pipe cactuses. Loss of these towering cacti, which can live one hundred to two hundred years, would, like the loss of the agave, predestine further declines in desert animal diversity.[7]

A similar situation can be found on many Pacific islands such as Samoa, where fruit-eating bats known as flying foxes (because of their fox-like faces) appear to be both the most influential pollinators and the chief seed dispersers for island plants. Like the bats of the Sonoran Desert, flying foxes are declining, in this case because of habitat loss, over-hunting, and exotic predators. The repercussions for the islands' plant communities and the animals that depend on them are equally worrisome.[8,9]

In Europe, the reproductive success of certain wildflowers already seems to be diminishing, thanks to the loss of butterflies and other insect pollinators. Researcher Ola Jennersten of Uppsala University focused on maiden pinks, a rosy-flowered, sun-loving *Dianthus* species. In fragmented habitats in southwestern Sweden, where patches of pinks are surrounded by agricultural fields, Jennersten found butterflies visited the flowers more rarely, and they produced significantly fewer seeds than those growing in undisturbed forest and meadow areas nearby. In fact, so few butterflies visited the pinks in those habitat fragments that they set no more seeds than plants that were covered with mesh bags and forced to self-pollinate. Jennersten suggests that most of the seeds produced in the fragmented area actually resulted from self-pollination.

Even though pinks and many other plants can self-pollinate, scientists worry that a high level of "selfing" season after season will eventually reduce genetic variation and perhaps make these species more vulnerable to stresses, such as disease or climate change, that could push them to extinction.[10]

The capacity for self-pollination represents one of several fail-safes that protect some plants when pollinator or seed dispersal services fail in any given season. Some plants can reproduce asexually or vegetatively—what gardeners will recognize as propagation from cuttings or root divisions—as well as by seed. Also, most plants are adapted to rely on a wide array of pollinators, or even on wind pollination. It follows that the plants at greatest risk of extinction are those with none of these backup mechanisms and tight mutualisms with specific animals.

Although such interdependencies are probably fairly rare, some can be found in tropical forests; for example, partnerships between figs and fig wasps, or orchids and orchid bees. Among the highly diverse fynbos or scrubland flora of the Cape region of South Africa, William Bond of the University of Cape Town reports that "many of the most beautiful plants are pollinated by a sparse and unusual pollinator fauna of long tongued flies, oil collecting bees, monkey beetles, large carpenter bees and the butterfly *Meneris tulbaghiae*," which is strongly drawn to spectacular red flowers. As a consequence, Bond believes many of the Cape irises, amaryllis, orchids, and other rare plants "seem particularly vulnerable to pollinator failure."[11]

Plant Resources

It's not surprising that plants can play key roles in their communities, since plants form the foundation of the food chain. The term keystone mutualist was coined largely with plants in mind. These are plants that sustain pollinators, seed dispersers, and animals known as mobile links through seasons of scarcity by supplying uniquely timed food resources. In turn these animals foster the reproductive success of other plants—perhaps the majority—in a community.[12]

One example of a keystone mutualist is *Banksia prionotes*, a shrub in the Proteaceae. At certain seasons of the year, honeyeaters and other nectar-feeding birds in the scrublands of southwestern Australia must rely solely on *Banksia* flowers for nourishment. The rest of the year, these honeyeaters vector pollen for many other plants and play an important role in maintaining what is, along with South Africa's fynbos, one of the most diverse plant communities outside the tropics.[13] Eliminate *Banksia* and there go honeyeaters, followed by any number of plant species dependent on their services.

In the tropics, keystone plants are often those that provide fruit for important seed-dispersing animals. The relationship between fruiting trees and fruit-eating animals is called a mutualism because both parties benefit. Animals need the nourishment in the nutrient-packed pulp and seeds, and the plant needs to have its seeds carried to new sites, perhaps gaps in vegetation where seedlings stand a better chance of surviving. Of course, as animals eat fruit, many of the seeds get digested, destroyed, or regurgitated on the spot. But a fraction may be deposited later in dung a good distance away from the parent tree. This service is valuable enough that 50 to 90 percent of the canopy trees and almost all shrubs and smaller trees in the tropical forests of Central and South America bear fruits that are fleshy, brightly colored, or otherwise adapted to attract animals.

The brilliant red-coated fruits of the tree *Casearia corymbosa* provide an example of this type of keystone mutualism. During a critical period each December when fruit is otherwise scarce in the forest, these fruits sustain twenty-two species of birds, including cotingas, toucans, parrots, and robin-sized tityras. If *Casearia* trees were to disappear from the La Selva biological preserve in eastern Costa Rica, Henry Howe of the University of Iowa predicts the loss would lead to the disappearance not only of masked tityras—the only birds that reliably spread the *Casearia's* seeds far and wide—but also of toucans. The toucans need *Casearia* to sustain them through December's scarcity, but by January they move on to other preferred fruits, ranging widely and scattering the seeds of a number of plants, including nutmegs. Thus the loss of *Casearia,* especially from a patch of forest isolated amid a landscape of agricultural fields and grasslands, could precipitate a slow-moving wave of extinctions among an array of forest species.[14]

The idea that seasonal famines could threaten birds and mammals in a habitat as seemingly lush as the tropical forest might appear counterintuitive. Yet John Terborgh of Duke University points out that the tropical forest is not, as the traditional jungle stereotype holds, "a Garden of

Eden offering a benign and constant climate and year-around abundance of fruit and other plant resources." Fruit production can be unpredictable and irregular, creating a life of feast or famine for the vertebrate animals that rely on it. Yet in some tropical forests, fruit-eaters comprise three-fourths of the bird and mammal biomass.

Why such skewed numbers? For one thing, the foliage of the tropical forests is laced with defensive chemicals and so is neither as edible as the grasses and forbs of the savannas, nor—given its treetop location—as accessible. Second, the plants needing seed-dispersal services have evolved fruits that are highly attractive to animals. Still, the volume of flowers, fruits, and other reproductive parts a forest plant can supply is small relative to its foliage, so a community dominated by fruit-eaters will contain much less animal biomass than a community of grazers. While the African grasslands can support up to twelve tons of vertebrate grazers per square kilometer, Terborgh found the forests around the Cocha Cashu Biological Station at the base of the Andes in the Peruvian Amazon can support only 1.6 tons of fruit-eating birds, bats, monkeys, squirrel-sized tamarins, agoutis, and other mammals per square kilometer.

The dependence of animals on fruit-producers highlights the importance of protecting certain keystone plant species in the tropics. At Cocha Cashu in Peru and also at Barro Colorado Island, Panama, Terborgh found that soft fruits are scarce during a three-month period from May to July. During that time, the monkeys, tamarins, and many other fruit consumers must rely almost completely on palm nuts, strangler figs, and nectar produced by a dozen keystone plants. These twelve plants represent only a fraction of the approximately two thousand plant species at the site in Peru, and yet they largely determine the animal carrying capacity—the number of animals that can make a living—in that forest.[15]

Further north along the Andes, in the mist-shrouded cloud forests of Colombia's central mountain region, Gustavo Kattan of the University of Florida has found that nearly two-thirds of the birds (168 species) are threatened with extinction if the extensive deforestation now underway continues. Most of these species are rendered vulnerable by their dependence on forest habitat. Prominent among them are the large fruit consumers, including parrots, toucans, and cotingas. Unless enough forest is preserved to support fruit-bearing plants in sufficient numbers to sustain viable populations of toucans and other fruit-eaters through seasons of scarcity, the integrity of the forest community may be jeopardized.[16]

Unfortunately, the repercussions of deforestation in Central and South America will not be confined to those regions, but will be apparent a continent away, in the backyards, parks, and woodlands of North America. Ongoing habitat fragmentation in these temperate regions will, in turn, make its mark in the faraway tropics. The connecting links are some two hundred-fifty species of birds known as neotropical migrants. They nest in North America each spring and summer, mostly in the eastern deciduous forests, then fly south to winter in tropical regions. In recent years, surveys in North America have found significant declines in half of those species, including various thrushes, warblers, vireos, cuckoos, flycatchers, and tanagers. The birds face threats in both their winter and summer haunts. Terborgh points out that forest cover in the eastern United States has declined by 40 percent since European colonists arrived, and much of what remains is fragmented second growth. Egg and nest predation is high in the fragmented areas, and populations of brown-headed cowbirds, which lay eggs in other birds' nests and usually doom the nesters' own young, have skyrocketed.

Terborgh also points out that more than half of these migrants depend on wintering quarters in southern Mexico, the Bahamas, Cuba, and Hispaniola, all areas that have experienced acute deforestation.[17]

What functional impact might the continuing decline of neotropical migrants have in North America? Most of these birds are insectivores, at least during their northern stay, and they consume an array of woodland pests such as spruce budworms, gypsy moths, and tent caterpillars. It seems fair to speculate that their loss might lead to more frequent outbreaks of insect pests. Indeed, when researchers in the Ozark woodlands of Missouri enclosed white oak saplings in nylon gill netting to exclude birds, the trees were attacked by twice the number of caterpillars and lost twice as much leaf area as uncaged trees nearby, where insect-eating birds foraged. By the second year of the experiment, the caged trees also showed reduced growth. These findings led Robert Marquis of the University of Missouri–St. Louis and Christopher Whelan of the Morton Arboretum in Lisle, Illinois, to predict that declines in insect-feeding birds, both neotropical migrants and resident songbirds, could reduce the productivity of North American forests.[18] Through habitat destruction, today's societies may simply be reenacting, unintentionally and ever so slowly, the Chinese sparrow massacre.

The problems faced by migrant species are not limited to the New World. Hundreds of so-called palearctic migrants move between Europe and the tropical regions of Asia and Africa. These birds also find themselves at risk at both ends of the journey, as well as during their travels. The once familiar white storks of Europe are now hard-put to find marshes where they can feed on frogs and other aquatic life, and they face a gauntlet of hunters as they brave their ancestral flyways to reach wintering grounds in sub-Saharan Africa. Meanwhile, drought and desertification in the Sahel have helped drive a 75 percent decline in whitethroat warblers that arrive in England to breed and feast on insects each spring.[19]

Seed Eaters and Grazers

As just described, destruction of critical plant resources poses a major threat to the complex web of plant–animal dependencies. Eliminating certain animals, directly or indirectly, can also be disruptive to a community over time.

Interestingly, the distribution of plants in both the temperate and the tropical regions of the Americas today may be quite different than when humans arrived on the continents at the end of the last ice age ten thousand years ago. The change may reflect the fact that these plant communities have been largely bereft of certain animal services since the demise of the huge fruit and foliage-eating Pleistocene megafauna—some believe at the hands of spear-throwing humans. The lowland forests of Central America, for instance, lost mastodon-like gomphotheres, giant ground sloths, glyptodonts, native horses, and other massive consumers of fruit, seeds, and foliage.

Daniel Janzen of the University of Pennsylvania and Paul Martin of the University of Arizona have suggested that plants that did not go extinct along with their giant seed dispersers still bear "puzzling fruit and seed traits" that can best be understood as living anachronisms. These are traits shaped by evolutionary interactions with long-gone animals. They include large, rich fruits that don't burst open when they ripen, hard-coated seeds that few animals in the forests today can crack, and an offering so plentiful in certain seasons that contemporary fruit-eaters cannot devour it all. Indeed, a high proportion of this type of fruit crop rots on the branch or on the ground below the tree, even in parks and protected areas where vertebrate fruit-eaters, from agoutis to monkeys to toucans, are abundant.

Without enormous beasts to swallow fruits by the hundreds and scatter piles of seed-rich dung across wide expanses of the landscape, researchers think many of the large-fruited trees, including the forest palms that Terborgh lists as keystone resources today, have lost range over the millennia. Janzen and Martin believe much the same thing occurred when temperate North America lost its fruit-eating megafauna; trees bearing large, sweet-fleshed fruits like the Kentucky coffee bean, honey locust, osage orange, pawpaw, and persimmon grew ever more sparse and limited in range.[20] (The loss of giant grazers, such as mastodons, may have altered the North American landscape even more profoundly, as you'll see in Chapter Seven.)

If toucans, agoutis, monkeys, and other contemporary frugivores and seed dispersers are allowed to go extinct, future naturalists may find themselves puzzling over seemingly ill-adapted traits in exceedingly rare stands of fruiting trees—the baggage of lost relationships today's human societies did not have the foresight to preserve.

Already human activities appear to be altering these mutualistic relationships indirectly, by disrupting the balance of animal seed-eaters in some tropical forests. The changes are not as dramatic as the alleged Ice Age massacre of mastodons and gomphotheres across the Americas, but they also may rearrange the mix of species in the forest community over time. Whether the changing biological landscape will still function in much the same way—and still sustain human welfare and economic activities—ecologists cannot predict.

The cascade of effects begins with the elimination of big cats, the jaguars and pumas that are the top predators in the neotropical forests. In the relatively undisturbed Amazonian forest around Cocha Cashu in Peru, jaguars and pumas prey on mammals they come across on the forest floor: peccaries, pig-sized rodents called capybaras and pacas, smaller agoutis, and raccoon-related coatimundis. This bounty supports high enough cat numbers to keep the populations of the prey species in constant check. Except for places like Corcovado National Park in Costa Rica, such intact neotropical systems have all but vanished in Central America and Mexico.

Even well-protected research sites have their problems. Barro Colorado Island, created when it was cut off from the mainland during the building of the Panama Canal, was too small to sustain large cats, which died out on the island. Without predators, small mammals have expanded to stunning densities. The island holds ten times the numbers of agoutis, pacas, and coatis as a similar area at Cocha Cashu and elevated levels of several other species such as peccaries and opossums. This population explosion has sent ripples through the island community.

For one thing, the increase in coatis and other nest predators is apparently responsible for the local extinction of many ground-nesting birds. Peccaries, pacas, and agoutis now devour so many fallen seeds from large canopy trees that the plant assemblage on the island may begin to shift toward smaller trees and shrubs. Such a scenario finds support from two sources. First, researchers have found a ten-fold increase in the survival of seeds and seedlings of two rain-forest canopy trees on the nearby mainland, where human hunting keeps mammal populations low in comparison with Barro Colorado Island. Second, nearby islands too small to support peccaries, pacas, or agoutis are dominated by a few large-seeded trees.[21]

Terborgh sees these suggestive findings as demonstration of the power of indirect effects. He writes, "If indeed jaguars and pumas limit the numbers of many terrestrial mammals, and if some of these, in turn, are important predators of bird nests, small vertebrates, seeds, and seedlings, we shall have to recognize that, in order to preserve diversity in tropical forests, we will have to maintain a more or less natural balance between top predators and their prey. Disrupting this balance—by persecuting top predators; by overhunting pacas, agoutis, peccaries, and perhaps other game species; or by fragmenting the landscape into patches too small to sustain the whole interlocking system—could lead to a gradual and perhaps irreversible erosion of diversity at all levels."[22]

The power of indirect effects on the plant–animal web also shows itself when the animals that disappear—through overhunting or habitat destruction—feed on foliage as well as fruits or seeds. Such is the case at the Los Tuxtlas Tropical Research Station in the Mexican state of Veracruz, at the northern-most limit of the neotropical rain forest. Evidence indicates that over a twenty-year period, the area's formerly sizable populations of peccaries, tapirs, coatis, white-tailed deer, and other browsers have been eliminated by hunting, illegal trade in animals and pelts, pet collection, and habitat destruction. When Rodolfo Dirzo and Alvaro Miranda of the Universidad Nacional Autonoma de Mexico examined seedlings and herbs in the lush understory of the Los Tuxtlas forest, the only leaf damage they could find had been inflicted by herbivorous insects. Nowhere could they find signs that mammals had nibbled, trampled, or pulled at the greenery.

To get some idea of the consequences for the forest, Dirzo and Miranda compared their site with another in the Montes Azules Biosphere Reserve in Chiapas, part of the largest remaining tract of tropical rain forest in Mexico. There they found animals in abundance, and a full 29 percent of the plants showed signs of vertebrate animal browsing. Yet, rather than exerting a negative impact on biodiversity, the browsers seem to have boosted plant diversity: The understory vegetation at Montes Azules contains three times as many species as Los Tuxtlas.[23,24]

How is it that browsers can contribute to the diversity of plants on which they feed? Such a relationship, it turns out, is not straightforward, but seems to depend both on the intensity of the grazing and the competitive abilities of the preferred food species—whether or not that plant has the power to monopolize the community if left unchecked.[25,26]

Many other examples of disruptions in plant–animal relationships have been identified, and they are not limited to the tropics. In the eastern Canadian Arctic, human activities are driving a population boom among lesser snow geese, voracious grazers that nest in the lowlands around Hudson Bay. The increase in goose numbers is attributed to reduced hunting pressure, more refuges along the migratory route, and the expansion of corn, rice, and wheat fields along the flyway. Between 1968 and 1991, for instance, the number of geese summering at La Perouse Bay, Manitoba, rose from eighteen hundred to twenty-three thousand pairs. The numbers alone were enough to stress food resources around the bay, leaving the ecosystem vulnerable. Then, during the 1980s, a series of late springs and hot summers conspired with the pressure from the geese to dramatically alter the coastal plant community.

Hungry geese arriving at the bay just as the late snows melted, but before plants had put up shoots, impatiently began grubbing for roots and rhizomes of grassy plants, creating vast patches of barren ground. The process doesn't stop with the grubbing, however. Intense summer heat on the unprotected soil increases evaporation and pulls up inorganic salts from the underlying marine clay layer to create hypersaline conditions that reduce the growth of forage plants. The end result is an expanse of barren mudflats along the coast, marked by the destruction of fully half the grassy salt marsh swards since 1985.

The geese themselves haven't fared well during all this. Underfed goslings have grown into puny geese, and today's adults are 13 percent smaller than the geese of 1969. But their destruction of the coastal vegetation has also had a snowball effect on other creatures, including a decline in shorebird populations and also in grazing ducks, such as widgeons.[27]

Invaders

Few regions in the world today remain untouched by the pervasive scrambling of the earth's biota that has accompanied five hundred years of steadily accelerating human commerce and travel. Immigration by animals and plants is nothing new, of course. There have always been wanderers, windborne vagrants, and stowaways on storm-tossed driftwood that reached even the most remote outposts. A small percentage of these immigrants survived, persisted, and spread, providing the raw material for the evolution of new, often uniquely local species exemplified by the biological wonders of island archipelagos like Hawaii and the Galapagos. But comparing that natural process with the flood of exotic plants and animals being let loose on native communities today is as misleading as comparing the background rate of species extinction with the massive biodiversity loss currently underway. The volume and pace of human-driven invasions and extinctions make both occurrences novel in the history of life.[28]

Many invasive organisms, such as rats, thistles, and fire ants, have been redistributed around the globe quite unintentionally, hitchhiking on transoceanic shipments of produce or even on the shoes or in the luggage of travelers. Zebra mussels and other aquatic invaders reached the North American Great Lakes in the ballast tanks of ships that were emptied upon reaching port.

Many other species have been deliberately imported as crops, garden plants, pets, fish farm stocks, and livestock. Too often these escape or are deliberately turned loose into native communities. At the height of European colonialism, in fact, importing and releasing novel species was enthusiastically pursued in the belief that native biotas needed "enrichment," a tenacious philosophy that brought starlings and European carp to the United States and filled the highlands of eastern and southern Africa with pine trees and trout. Many, perhaps most, ecologists today subscribe to the notion that all such non-native species released into the environment are weeds at best, biological pollutants at worst.[29]

Certainly, the United States has received its share of troublesome additions: gypsy moths, kudzu vines, Dutch elm disease, chestnut blight, water hyacinths, dandelions, cheatgrass, zebra mussels, brown rats, and house mice. Yet in most parts of the world, people express a fondness for the rare and unusual, a sentiment that fuels a lucrative international trade in orchids, bulbs, boutique fruits and vegetables, exotic fish and game animals, parrots, iguanas, llamas, and other novel pets and livestock. Modern agriculture prospers on the wide dissemination of new crop varieties. How many Americans would be willing to give up honeybees, wheat, apple trees, wine grapes, cattle, ring-necked pheasants, or even the feral mustangs and tumbleweeds (Russian thistle) woven into the lore of the American West? Even the seemingly natural ecosystems around us are so laced with invaders that people often don't even realize who belongs. The concept of belonging, of course, is weighted with value judgments. That's because decades or even centuries after their arrival, many exotics have become integral parts of the landscape, culturally and sometimes ecologically.

An example can be found in the tale of opossum shrimp and kokanee salmon with which this chapter opened. The kokanee were no more native than the opossum shrimp. Stocked into Flathead Lake in 1916, the kokanee quickly displaced native cutthroat trout as the dominant sport fish. After nearly three human generations, people have come to regard the kokanee as belonging in the lake. They are not completely wrong. The paradox is that, over time, the kokanee has developed into an ecologically important player in the highly modified and biologically impoverished environment the kokanee helped to create in the American West.

A century ago, for example, bald eagles migrating south from the Canadian prairies could have fortified themselves on sea-run salmon returning to the Columbia and Snake River headwaters to the west of Glacier or scavenged bison carcasses on the plains to the east. By the time the kokanee were introduced, the bison had been hunted to near-extinction, and these days only a handful of salmon make it back through the gauntlet of dams to Idaho waters. For decades the kokanee provided a substitute for some of these diminishing resources. The collapse of the kokanee population must be viewed as another lost link in a highly changed, fragile food web that still sustains some of North America's most revered wildlife.[30]

Very few introduced species ever attain such ecological significance, largely because most new arrivals simply fail to establish themselves and survive. Some do successfully establish, yet cause little change in their new communities. Others establish and quickly wreak havoc among native species, preying on them, out competing them for space or food, spreading diseases or parasites among them. A few of these invaders—perhaps 10 percent according to an earlier SCOPE study—have impacts strong enough to alter larger-scale ecosystem processes.[31]

But which species will make that 10 percent? So far, predicting in advance which species will be successful invaders and which communities are most vulnerable to invasion is a very inexact science. The best definition of a good invader is somewhat circular: A good weed is weedy; that is, it colonizes quickly, grows and reproduces rapidly, thrives on disturbance, and is not fussy about its nourishment. Peter Vitousek of Stanford suggests that among the successful weeds, the ones most likely to make their mark on ecosystem processes either (1) differ substantially from native species in the way they acquire or use resources such as nutrients or water; (2) change the food web; or (3) alter the intensity or frequency of disturbances, such as fire.[32]

As for the target communities themselves, species-rich assemblages generally appear to be more resistant to disruptions by new arrivals than species-poor ones. Yet disturbances seem to make any community more susceptible to invaders by stressing or eliminating native species and freeing up resources. Human disturbances, such as road building, clearcutting, and plowing, often involve increased traffic in men and machines, too, providing more opportunities for invaders to catch a ride.[33]

Although most new arrivals fail to gain a toehold, many deliberate introductions of fish into freshwater lakes and streams have been successful. Why? Because the prevailing ethic in fisheries management until recently has emphasized repeated stocking of favored game fish and commercial species to "improve" local fisheries—in other words, to make fishermen happy. In many regions of the world, in fact, that ethic still has a firm grip on fisheries managers. Fishermen themselves occasionally perform illicit bait-bucket introductions in their favorite waters.[34]

Just such a stocking apparently took place in recent years in Yellowstone National Park, an act that park officials have declared environmental vandalism. The act became apparent in the summer of 1994, when a few fishermen pulled exotic lake trout from Yellowstone Lake. When the surprised anglers reported their catch to authorities, alarmed

scientists placed test nets around the lake. From their catch, they estimated that the population of these invasive predators had already grown
into the thousands, and perhaps tens of thousands. Yellowstone Lake
and its tributaries support one of the last thriving native cutthroat trout
fisheries in inland North America, a unique resource in a spectacular setting that has become one of the world's prime trout-fishing destinations.
Biologists are deeply worried because lake trout prey on the smaller cut-
throats and could deplete their stocks by as much as 80 to 90 percent.

The consequences for other species in the area also could be severe.
In early spring, the cutthroat move into Yellowstone Lake's feeder
streams to spawn, supplying a critical food resource for hungry grizzly
bears emerging from hibernation, bald eagles, and forty other native
species, from otters and minks to osprey and loons. In contrast, lake
trout spawn in much deeper water where few fish-eating birds or mammals can nab them and a limited group of human fishermen are interested in pursuing them.[35]

If aquatic food chains seem especially vulnerable to exotic predators,
island communities are even more so. That's because most islands have
simplified food webs that often lack top predators or large herbivores.
Some of the most devastating invasions have involved releases of rats,
cats, mongooses, rabbits, snakes, pigs, goats, and deer on oceanic islands where they have no natural enemies to constrain them and where
the local organisms have evolved no defenses against them. Plants with
no thorns or chemical defenses, and flightless or ground-nesting birds
have proven especially vulnerable.

Goats introduced by early seafarers, for instance, have devastated vegetation on many Pacific islands, from the Galapagos north to the Channel Islands. Browsing goats and rooting pigs are major threats to the native flora and fauna of Hawaii. Small wonder that more than a third of
the officially threatened or endangered species in the United States live
on the Hawaiian Islands. Goats and pigs trample and devour the islands'
plants, denude slopes, increase soil erosion, and spread the seeds of exotic plants in their dung, encouraging further incursions by exotic banana poka vines, Himalayan ginger, and Brazilian strawberry guava
trees. Invaders like these are so ubiquitous now that most visitors to
Hawaii assume they are native.[36]

Getting rid of an alien species once established has proven almost impossible, and the cost of minimizing the ecological damage can mount
enormously over time. The U.S. National Park Service estimates it will

spend $300,000 a year just to suppress—not eliminate—invasive lake trout in Yellowstone Lake.[37] The cost of controlling predatory sea lampreys in North America's upper Great Lakes over three decades has been enormous, and there is no end in sight. Starting in 1962, fisheries officials began using selective chemicals to kill off sea lamprey larvae in spawning streams.[38] By 1989, lamprey control had cost the United States and Canada $127 million. The program continues at a cost of $10 million a year to protect the lakes' reconstituted sport fishery. That billion-dollar fishery is maintained largely by stocking of hatchery salmon and trout at a cost of another $10 million a year, an operation that will also have to be continued indefinitely because these fish have achieved only limited natural reproduction in the lakes.[39]

Pests, Parasites, and Pathogens

Some of the most dramatic disruptions to a community are caused by invasions of exotic pests, parasites, and pathogens. Every natural community, of course, already carries its own burden of parasites and disease-causing organisms. In North America, for instance, each tree species on average hosts fifteen fungal pathogens, each shrub seven, each herbaceous plant five. These cause aptly named rusts, smuts, wilts, rots, yellows, diebacks, damping off, and mildews. Virus loads are less well studied, but forest trees may host anywhere from one to seven disease-causing viruses per species. Then there are some three thousand species of parasitic plants, such as mistletoes, as well as untold numbers of troublesome soil nematodes and bacterial plant pathogens.[40]

Disease outbreaks are certainly part of the natural cycle. Occurring periodically in virtually all populations, they can weed out less fit or vigorous individuals, recycle essential nutrients, and influence the mix of species and the direction and pace of successional change in a community.[41]

A case in point is the fungal root-rot dieback that sweeps through stands of mature mountain hemlock every 90 to 150 years in the conifer forests of the U.S. Pacific Northwest. The outbreak starts from one or a few infected trees and moves out in a radial pattern as the fungal mycelia spread underground from the roots of one tree to another. Pamela Matson of the University of California, Berkeley, and Richard Boone, then of Oregon State University, found that, after the disease front has moved through a stand, nitrogen release in the soil doubles.

The standing dead trees often give way to fire and then to young hemlock seedlings that grow in the fire-swept clearings, where the extra sunlight and nitrogen help them withstand the effects of the fungus. The young hemlocks are subsequently joined by resistant lodgepole pines and an array of other species, depending on the soil moisture, creating a vigorous and often more species-rich patch of young forest.[42]

However, when pathogens invade new regions, where local species have developed no resistance, they can push a community so far that it may never be able to return to its former state. Shortly after the start of the twentieth century, the fungus that causes chestnut blight spread from New York City through the eastern forests of North America, eliminating 3.6 million hectares of magnificent chestnut trees within fifty years.[43] Although other trees have proliferated to fill in for the once-dominant chestnuts, the loss has permanently changed the look of eastern deciduous forests. The impacts on forest processes appear to have been minimal so far, although no before-and-after data exist to judge impacts on soil, mast production, and other functions.

The same cannot be said for the forests of Australia and Tasmania where the assault of another invading fungal pathogen named *Phytophthora cinnamomi* has precipitated a cascade of ecological degradation. In its endemic setting of South Africa, this particular root-rot fungus is fairly benign, killing only scattered individuals in patches of natural vegetation. Where the fungus has been introduced, however, it has been extraordinarily aggressive, capable of infecting more than one thousand plant species, especially woody perennials. It was introduced to the eucalyptus forests of Victoria and Western Australia about the same time the chestnut blight fungus appeared in New York.[44] As it spread through Australian forests on the tires of road construction vehicles, the fungus began destroying vast tracts of species-rich heath and eucalyptus woodlands, sometimes converting the entire assemblage into open tracts of sedge. In severely infested areas of the Stirling Range National Park in southwestern Australia, studies showed that 85 percent of tested species in the protea family were susceptible to *P. cinnamomi,* including the majority of those with vertebrate-pollinated flowers. This raises the specter of a slow cascade of indirect effects precipitated by the loss of nectar and also fruit and seed resources that sustain a variety of birds, mammals, and insects.[45]

The fungus has apparently arrived more recently in the temperate rain forests of western Tasmania. There nearly one-third of the plant species tested have proven vulnerable to *P. cinnamomi,* including half the woody species. The fungus has invaded forests along road tracks and in places where the canopy has been destroyed by fire, allowing the sun to warm the soil and create hospitable conditions for the pathogen. Such disturbances have been increasing, of course, as the forest is exploited for mining, timber, tourism, and hydroelectric projects.[46]

Yet another fungal disease, pine blister rust, has altered the mix of species in many coniferous forests since it was introduced to the Pacific Northwest from Europe in 1910. In the Northern Rockies, the slow elimination of whitebark pine by blister rust, combined with fire suppression, has led to unaccustomed domination of many forests by Douglas fir and grand fir. These species, in turn, are susceptible to *Armillaria,* a native fungal disease that has little impact on whitebark pine but is now thriving on the ascendance of the firs.[47]

Other, more cryptic ecological effects may be underway, too. Whitebark pine are hardy, long-lived trees that can pioneer cold, windswept zones at the upper limit of the alpine timberline where no other trees can survive. These pines shade the slopes and slow the spring snowmelt, helping to prevent flooding and assure a steady flow of water in the streams late into the summer—a service that is being lost as the pines die and the treeline moves lower on the mountains. This change may have serious consequences for water supplies in the Rockies, northern Cascades, and Sierra Nevada ranges and eventually for grizzly bear populations and other wildlife.

Whitebark pines produce unusually large, nutritious pine nuts prized by bears, squirrels, birds like Clark's nutcracker, and other wildlife. In both Glacier and Yellowstone National Parks, these nuts are a critical, often predominant, food for bears in the fall as they raid squirrel and nutcracker caches to fatten up for hibernation. Yet across western Montana, including the Glacier area, 42 percent of whitebark pines have been killed by blister rust in the past two decades and 89 percent of the rest are infected. Scientists so far offer little hope of halting the spread of blister rust. Early and largely unsuccessful disease control efforts were abandoned in the 1970s. Even if a few resistant trees survive and reproduce, a whitebark pine must be about a century old before it can produce a significant nut crop. To find alternate foods, grizzly bears must forage at lower elevations, a circumstance that brings them into more frequent conflict with humans and inevitably results in more bear deaths.[48,49]

Novel disease organisms that attack animals can also generate indirect effects that severely impact human societies as well. The classic example is rinderpest—horned beast plague—apparently brought into the Horn of Africa from India in 1889 by infected cattle. The disease spread quickly. A letter to the Imperial British East Africa Company in 1891 reported that in some regions of Uganda, fifteen hundred kilometers to the southwest, "almost all the game, including the small antelope, seem to have perished." Within six or seven years, the viral disease had swept through tropical Africa all the way to the Cape. Along the way, populations of large grazers, such as wildebeest and buffalo, were decimated, as were domestic cattle herds. Pastoral people endured severe famine. Loss of grazing animals and complex changes in the frequency and intensity of fires caused vast stretches of open savanna to revert to bush and woodland, where tsetse flies, notorious carriers of sleeping sickness, thrive.

Rinderpest was largely brought under control by cattle vaccination programs in the mid-1960s, but in recent years, humans and their domestic animals have apparently loosed a new round of viral diseases into wild populations in Africa. By the 1970s, researchers in Tanzania's Serengeti Plains were reporting occasional outbreaks of canine distemper among wild dog populations, apparently acquired from contact with domestic dogs kept by Masai herdsmen in the region.[50,51]

Canine distemper, like rinderpest, is one of the morbilliviruses, a group that includes human measles. In the 1980s and 1990s there have been several reports of new morbilliviruses, or new variants of old ones, killing dolphins and other marine mammals in the Mediterranean and the Atlantic, seals in Siberia's Lake Baikal, and lions, hyenas, leopards, and foxes in the Serengeti. The seals are believed to have been exposed after carcasses of sled dogs killed in a distemper outbreak were dumped into the lake. In the Serengeti, a distemper epidemic reported to have killed one thousand of the park's three thousand lions during 1994 has been traced to a new variant of the canine virus that swept through the dog population in nearby villages during 1993 and 1994. [52,53]

Avian malaria and the *Culex* mosquitoes that carry it were introduced to Hawaii and other islands in the early nineteenth century, probably by whaling ships. The disease has since decimated many native bird populations that had no natural resistance to the microbe. Surviving Hawaiian birds like the brilliant orange-red nectar-sipping 'i'iwi have largely retreated to the malaria-free zones above one thousand meters, leaving resistant exotic birds to take over below.[54]

While it is impossible to forecast how losses of unique island birds—

or marine mammals, or top predators from the African savannas—will impact ecological processes, it's easy to calculate direct impacts on the human enterprise from loss of some less charismatic creatures. The honeybees of North America are a case in point; their current plight parallels that of the kokanee salmon. Honeybees, like the kokanee, are exotics in the United States. Introduced from Europe by early colonists, honeybees have largely displaced the native bumblebee and, as a result, have developed into ecologically indispensable workers. Each year they now pollinate about $10 billion worth of crops, from New York pumpkins to California almonds. Honeybees, both feral colonies and those maintained by beekeepers, have been hit hard since the mid-1980s by the invasion of two deadly parasitic mites. The result has been serious pollinator shortages in some farming regions.[55]

Pest Control Providers

As the case of the honey bee illustrates, economic losses can be severe when introduced pests or native pests released from natural controls invade our highly simplified and unstable agricultural ecosystems. By one account, the world's croplands support some fifty thousand plant pathogens, nine thousand insects and mites, and eight thousand weeds. Depending on the crop and the region, from 20 to 70 percent of those pests could be exotics.[56]

The loss of productivity in these managed ecosystems can be staggering. Globally, farmers lose 30 to 40 percent of preharvest crop yields to pests and disease.[57] In the United States alone, crop yields are reduced by about one-third by pests: 160 species of bacteria, 250 viruses, 8,000 pathogenic fungi, 8,000 insects, and 2,000 weeds.[58]

When the natural enemies that provide checks on crop-destroying pests are considered, the number of species involved in either pestilence or pest control skyrockets. Indeed, insects involved in some sort of biocontrol work as plant-feeders (which may eat crops or check weeds), predators of plant-eating insects, parasites, or pathogens, may constitute more than half of all species on the earth.[59] This tally doesn't include insectivorous birds and other animals that supply pest-control services.

One has only to remember the insect outbreaks that followed the Chinese sparrow massacre to realize the value of natural pest-control services. Yet China's experience is replicated time and again throughout the world when heavy applications of pesticide are used without regard for beneficial insects that are keeping target and other pests in check. Extermination of natural enemies often releases minor pests from control, allowing their populations to explode and promoting them to nuisance status.

The introduction of modern rice varieties across Asia, starting in the late 1960s, was accompanied by increasing insect damage. Leaffolders, caseworms, armyworms, and cutworms, which had once been relatively minor pests in paddies, began to proliferate. Farmers turned to pesticides to protect their new crops, but repeated spraying created a new menace by killing off the natural enemies of the brown planthopper. Populations of planthoppers have been building since then, periodically causing outbreaks of "hopperburn" damage in Thailand and elsewhere. In Indonesia, planthoppers destroyed $1.5 billion in rice during a two-year period in the late 1980s, an event that caused the Indonesian government to end its subsidies for pesticide purchases and begin to steer farmers toward less destructive forms of pest control.[60,61]

A legendary scourge of the southern United States is the cotton boll weevil, a species agricultural officials have been battling through aerial spraying since 1974. Farmers in the Rio Grande Valley of Texas, the nation's largest cotton-producing state, voted in 1994 to join in the eradication program, a decision they came to regret after a single season. After aerial spraying with malathion, which officials say killed 98 percent of the weevils, pests like beet armyworms and aphids exploded. The cotton crop dropped from 308,000 bales in 1994 to just 54,000 in 1995, one of the worst harvests in the history of the valley. Farmers blame the boll weevil spraying for killing off spiders, wasps, and other natural enemies that usually kept the aphids, armyworms, and even the weevils at manageable levels. In early 1996, they voted to drop out of the program, and farmers in some other states were considering abandoning the spraying effort, too.[62]

Of course, misuse of pesticides and other farm chemicals isn't the only way to disrupt natural pest control services. Timber managers in the southern United States promoted fusiform rust "from a curiosity in 1930, to a major problem today in both plantations and natural forests" simply by altering the distribution of native pines. Longleaf pine, which is resistant to the rust, was logged extensively in the nineteenth century. Loblolly and slash pines, valued for their rapid early growth but more susceptible to rust, were replanted or allowed to recolonize clearcuts.

This switch, along with the suppression of fires, upset the ecological balance and created conditions conducive to rust disease.[63]

Another blow to beneficial insects comes when large sections of the landscape are converted to crop monocultures; the insects are left with no natural habitat. Practices such as rotating high-value crops like corn with cover crops of clover or alfalfa, intercropping or other use of multiple crop species, or leaving hedgerows or other natural vegetation alongside fields can help to sustain natural pest-control services even in highly modified systems. From the Dustbowl era of the 1930s until the 1950s, the U.S. Soil Conservation Service encouraged farmers to plant rose hedges along the contours of soil terraces to harbor predatory insects that help control crop pests. The practice fell by the wayside with the intensification of agriculture in the 1960s.

Hedgerows along the borders of fields in Bavaria date back hundreds of years. In a land where most forests have been converted to timber plantations, these hedgerows now represent Germany's most diverse woody vegetation, containing as many as thirty species of woody plants. They also serve as prime habitat for herbivorous insects, most of which are highly specialized feeders that have no interest in nearby crops. Instead, their presence attracts generalist predators and parasites, which not only feast on them but consume aphids in the neighboring grain fields. Because of these insects, northeast Bavaria is one of the few parts of Germany where farmers have no need to spray for wheat aphids.[64]

Devising new biological controls for pests after natural control services have been lost or destroyed, or pests have been transported to regions far from their natural enemies, has proven a costly, hit-or-miss effort. The results seldom equal the services provided by natural ecosystems. One hurdle has been scientists' limited understanding of what or who controls any given pest in its native setting; another is the inability to predict which potential control agents will invade successfully and go to work in a new setting.

David Pimentel of Cornell University points out that, on average, scientists introduce seven natural enemies of crop pests before finding one that's at least partially successful in a new setting. However, for the corn borer, some twenty-six natural enemies had to be imported before even limited control of this pest was achieved. Even tougher has been finding an effective natural enemy for the gypsy moth. Of forty that have been introduced into the forests of the northeastern United States so far, only ten survived, and only one of those does a creditable job of helping control this Eurasian invader. This one imported enemy, in concert with ninety native predators that attack the gypsy moth, still hasn't reined in the voracious caterpillars that annually cause tens of millions of dollars worth of tree damage.[65]

A far better approach for the future, of course, is to try to screen out potential pest species before they can gain a toehold. In 1990 the U.S. Forest Service took the unusual course of banning imports of raw logs from Siberia until a team of entomologists, disease specialists, and economists could evaluate the risks posed by possible hitchhikers. After assessing the potential mischief that might be caused, the team concluded that invasion by a single new forest pest, such as the spruce bark beetle, could cost the timber industry as much as $58 billion. Permits to import unfumigated logs were denied.[66]

Carriers of Human Disease

Few people realize the extent to which fraying the community web can affect human health. Those who might dismiss the loss of a few invertebrates as insignificant should consider the fact that ecological disruptions often release pests, pathogens, and carriers of human disease, sometimes with deadly consequences.

One dangerous disease vector that has hitchhiked to a world of new opportunities thanks to human commerce is the Asian tiger mosquito, *Aedes albopictus,* a native of Japan that first showed up in the United States in 1985, presumably aboard a shipment of used tires. By the early 1990s, the mosquito could be found throughout much of the Southeast and Midwest and was feared to be spreading. This pest has also invaded parts of Africa, New Zealand, Australia, South America, and southern Europe, perhaps by the same tactic. The health threat comes from the serious diseases this mosquito can carry, including various strains of encephalitis, yellow fever, and dengue or "breakbone" fever. The mosquito is already a suspect in yellow fever outbreaks that struck Brazil in 1986 and Nigeria in 1991, as well as in a series of encephalitis epidemics that hit Florida in the early 1990s. Public health officials are also worried that this new carrier may boost the worldwide resurgence of dengue fever, which is now epidemic in the Caribbean and has already shown up as sporadic outbreaks in Houston and other U.S. cities.[67,68]

Another Asian native, a strain of cholera bacteria, was carried to Peru in the bilge water of a Chinese freighter in 1991. The bacterium, released into Peruvian coastal waters, soon spread through the marine plankton community and into freshwater drinking supplies where it infected people directly. The cholera bug also infected fish, shrimp, and mussels, and many people who avoided contaminated water supplies were stricken after eating undercooked or raw seafood dishes, such as ceviche. Within two years, the cholera epidemic had struck a half-million people in the region.[69]

Even local pathogens can burgeon into major health problems when changes in land use allow their populations to explode, bringing them into unaccustomed contact with humans. These agents need not be infectious organisms. Artificial concentrations of a single species, such as monoculture plantings of crops or timber, can exaggerate natural but quite unpleasant biotic interactions, such as human pollen allergies. Consider Japan, where an aggressive fifty-year government effort has replaced 20 percent of the diverse native forests of oak, maple, and evergreen with monotonous but economically valuable plantations of cedar. The result has been an unprecedented level of misery for the human population. As waves of cedar pollen sweep over the country each spring, two huge electronic billboards in Tokyo flash the pollen count and wheezing residents call allergy hot lines. One in ten Japanese are now affected by pollenosis, a level of allergic reactions to cedar pollen unheard of in any other nation—or indeed in Japan for the thousands of years that cedars grew amid a variety of other species in natural forests.[70]

Sometimes human invasions help disrupt the links between disease agents and their usual hosts. Lyme disease, a spirochete infection carried by deer ticks, is a case in point. The disease was virtually unknown in humans until two trends converged in the 1970s: first, explosive growth among deer and mice populations, the natural hosts of the pathogen; and second, the push of suburban homes and human hikers into the forests of the northeastern United States. By entering the woods, humans are drawn into a complex disease cycle apparently paced by the boom-and-bust pattern of acorn production in the oak-dominated woodlands, a cascade of events worked out by ecologists Richard Ostfeld and Clive Jones of the Institute of Ecosystem Studies in southeastern New York, and Jerry Wolff of Oregon State University.

The cycle starts with acorns, which are a favorite food of both white-tailed deer and white-footed mice. Oaks only yield a bumper crop every two to six years. In the boom years, mice store a rich winter larder of acorns and unaccustomed numbers of deer move into the woods to feast. The deer, in turn, carry adult deer ticks, which drop off and lay eggs in the leaf litter. The following summer when the eggs hatch, the pinhead-size larvae discover a population explosion of mice. It is mice that carry the Lyme disease spirochete in their blood and pass the infection to tick larvae that bite them. The following spring, when the infected tick larvae molt into nymphs, they seek another blood meal from mammals entering the woods and their bite transmits the pathogen into the animal's bloodstream.[71]

The first human outbreak of Lyme disease was recognized in 1975, when fifty-one cases were reported from Old Lyme, Connecticut. Twenty years later, Lyme disease occurs throughout the United States, although it remains concentrated in the northeast. The disease is even more prevalent in Europe, especially Germany with its equally abundant deer populations, where more than thirty thousand people have been infected.[72]

A number of other emerging or resurgent disease problems worldwide are linked to land use changes that bring humans and disease vectors together. Throughout the tropics, artificial water impoundments, such as dams, irrigation works, and continually flooded rice paddies, increase breeding sites for mosquitoes, aquatic snails, and other disease vectors and often draw thousands of human laborers into new areas. For instance, intensification of wetland rice cultivation across Asia, including the introduction of short-cycle rice varieties that allow two to three crops a year in continuously flooded fields and require the heavy use of pesticides, has created conditions that favor snail-borne schistosomiasis and helped to drive an increase in encephalitis and malaria mortality worldwide.[73] More than thirty thousand farmers and field workers contract Japanese encephalitis each season in the flooded rice fields of Asia.

In the Argentine pampas, the conversion of rangelands to maize fields has enabled the formerly rare field mouse to expand. Unfortunately, the mouse is both a host and vector for the virus that causes Argentine hemorrhagic fever. As mouse populations in the fields have grown, more farmers have contracted the virus and passed it to family members. More than twenty thousand people have been infected since the virus was identified in 1958, and a third of these have died of hemorrhagic fever.[74]

Even in coastal waters, human activities can induce species changes that affect human health. Overfertilization of nearshore waters by human sewage effluents and industrial and agricultural runoff may promote what Theodore Smayda of the University of Rhode Island has called a global epidemic of novel phytoplankton blooms. Periodic or seasonal blooms of algae are natural phenomena in some regions. The Red Sea, for instance, takes its name from the color of the water during so-called red tide outbreaks. Many algae that bloom are benign, but others produce powerful toxins that kill fish and contaminate oyster and other shellfish beds, posing a danger to people who eat the shellfish. Although tracking the frequency of global algal blooms is difficult, scientists suspect that blooms of toxic and noxious algal species have increased in both frequency and intensity worldwide since the 1960s, at least partly because of nutrient enrichment of coastal waters. Between 1976 and 1986, for instance, Hong Kong's Tolo Harbor experienced an eight-fold increase in the number of red tide outbreaks, events that coincided with a six-fold increase in the human population and a two-and-a-half-fold rise in nutrient levels in harbor waters. Since the 1960s, an algal species seen previously only off the northeast coast of the United States has spread through North European waters, turning the surface yellow-brown and causing massive kills of sea trout and other fish during its frequent blooms. Another species known from southern California waters has spread to shellfish beds off Spain, Japan, and Tasmania.

Whether all of these toxic algal species are being spread in ballast waters or simply being promoted from rare to problem status in coastal communities where they have long existed is difficult to tell. Either way, the increase in fish kills and shellfish-borne diseases and poisonings among humans seems irrevocably linked to our own disruptions of coastal ecosystems.[75,76] It is but one more not-so-subtle signal of the ripple effects human activities are setting in motion in natural communities across the globe—ripples that may cost us dearly in lost ecological services and quality of life.

CHAPTER FOUR

Water: The Essence of Life

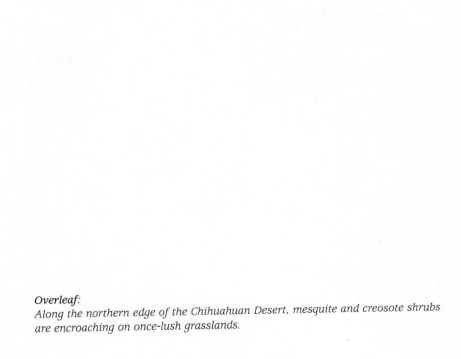

Overleaf:
Along the northern edge of the Chihuahuan Desert, mesquite and creosote shrubs are encroaching on once-lush grasslands.

*T*he earth possesses the only liquid water in the solar system. Indeed, it was a uniquely wet planet for a billion years before taking on its second distinction as the only living planet. Life sprang from the waters of ancient oceans, and the fate of living things remains intricately bound to processes of water. Salty or fresh, the water that bathes three-fourths of the earth's surface provides habitat to many organisms, and essential sustenance to all. Our Paleozoic ancestors crawled from the seas 400 million years ago, yet we and our fellow land creatures still carry within our tissues a remnant of those early oceans. Like most organisms, humans are two-thirds water, and we must consume roughly two liters a day to flush and replenish our vital processes. Externally, too, the continual renewal and cycling of water through plants and animals and across the landscape is critical to the productivity of the lands, lakes, and coastal waters from which we harvest our food.

Although water dominates the earth's surface, 97 percent of it is salty. All but a mere fraction of fresh water is locked away in glaciers, polar ice caps, and underground aquifers. The fresh water rivers, lakes, and reservoirs available to slake the thirst of terrestrial creatures, and the clouds overhead that promise rain on any given day, contain less than half of one percent of the planet's fresh water.[1] Even this water, though it may be temporarily impounded, moves ever downhill toward the sea.

Fortunately for life on earth, the ocean is a reservoir, not a grave for fresh water. Water cycles constantly between the atmosphere, land, and sea. It is breathed out pure and fresh from the evaporation of oceans, from the open waters and soils of the continents, and by the life processes of plants and animals, a great exhalation of water vapor powered by the energy of the sun. Aloft, this vapor cools and condenses into clouds, rains out, and makes its way back toward the sea.

On a global scale, this cycle is eternal and enduring. The flux was set in motion by physical forces long before the thin green shell of life took charge of the earth's surface. Yet living things can significantly modify the quality, speed, and flow of rainfall on its return journey to the sea. The most grandiose attempts by humans to modify the water cycle are dams, reservoirs, and diversion channels that have radically altered the quality and flow patterns of most of the earth's major rivers. The global climate changes that human civilization has unwittingly set in motion by unrestrained emissions of greenhouse gases may cause a major redistribution of the earth's rainfall.

Beyond that, human activities also modify the water cycle indirectly— usually in unwanted ways—by altering biological diversity. It is this link between water and the organisms that live in it or alter its composition and flow that forms the basis for this chapter. Traditionally, we humans have been blind to the role that organisms other than ourselves play in directing water processes. Their efforts seem dwarfed by our engineering feats. For people concerned with the degradation of our land and waters, how could the removal of trees or termites or fish rival the unintended ecological abuses wrought by major engineering works? For example, water is being pumped from aquifers faster than the rains can replenish it. As a result, Mexico City and numerous other urban areas are subsiding, literally sinking into the earth.[2] Reservoirs worldwide have drowned a total land area the size of France.[3] Many of the earth's great rivers like the Ganges, the Colorado, and the Nile send only a trickle of their former flows to the sea thanks to dams and diversions. Dams inundate upstream floodplains, destroy running-water habitats, block fish migrations, squander vast supplies of water to evaporation, and trap silt and nutrients that once flushed downriver to enrich the soils of deltas. Add the dumping of industrial wastes and sewage into open waters.[4] Could biological losses or species invasions possibly make things worse?

Perhaps the best way to appreciate the impact of living things on the water cycle is to follow the fate of rainfall and chronicle some of the ways in which biological changes, from losses of entire plant and animal communities to removal or additions of single species, alter water processes in undesirable ways. For most of the rain that falls, the earthly journey is short. Of the 113,500 cubic kilometers of rain that strike the land each year, two-thirds evaporates directly from wet leaves and soil or gets drawn up by plant roots and transpired back to the atmosphere, as crops, meadows, and forests grow. The remaining third either percolates through the soil into underground aquifers or runs off across the surface to join the oceanward flow, replenishing lakes, rivers, and other aquatic habitats along the way and creating a vast arterial system that connects alpine forests to coastal marshes.[5]

Yet rain is hardly benign. Left to beat down onto bare ground or run unchecked across the landscape, water is a powerful solvent that scours away soil and strips it of mineral nutrients that make it fertile. Plants and, later, animals had to devise strategies to blunt the erosive impact of rain, capture or redirect it, draw water and nutrients up into the food web, and slow the rush of water across the landscape.

This chapter begins by exploring the fate of rain in the dryland regions that cover a third of the earth's surface and by examining the link between plant cover, water cycling, and land degradation.

Waterlogging and Salt Scalding

The would-be farmers who settled Australia's southwestern frontier at the turn of the century certainly must have expected water problems, but never the kind their grandchildren face today. The region as they found it was an ancient expanse of rolling sand plains and uplands, devoid of standing water. Diverse mosaics of heath scrub and eucalyptus woodlands clung to the arid landscape. Today, more than 90 percent of that tough native scrub and woodland habitat is gone, cleared to make way for plowed fields and sheep pastures. This region has become one of Australia's premier farming areas, its 15 million hectares turning out $2.8 billion worth of wheat and wool each year.

Yet a cloud hangs over the future of this land and the people who depend on it for their livelihoods. Despite the development of better crop varieties and heavy investments in farm machinery, fertilizers, and weed and pest control, wheat yields have plateaued over the past twenty-five years. Scientists believe the potential gains from those investments have been offset by the massive deterioration of the land that has occurred after barely a century of cultivation, deterioration caused by, ironically, too much water.

More than half the arable land in the region suffers from some type of soil degradation that affects crop production, including salt scalding, waterlogging, erosion, and flooding. Salt intrusions have ruined 3 percent of the land, and agricultural officials fear 15 percent of the land could be lost to farming over the next thirty years. Increasing salinity could eventually threaten drinking water supplies for the city of Perth.[6]

Why did a water cycle that once supported one of the richest plant communities outside the tropics change so dramatically when the eucalyptuses, banksias, hakeas, and saltbush of the heath and woodland were replaced by wheat? Over the years, researchers have teased out increasingly detailed answers by examining the ingenious ways in which native plants deal with sparse rainfall.

One of those investigations began in the early 1980s, when a handful of eucalyptus trees in Wheelock's Catchment north of Newdegate township were fitted out with two-liter plastic bottles fastened around their stems and coffee-tin-sized plastic funnels dangling below their canopies. Robert Nulsen and his team from the Division of Resource Management in South Perth festooned the trees that way to find out exactly where the rainfall was going.

At the time, Wheelock's Catchment hosted a remnant of native mallee scrubland, a type of eucalyptus community, that formed a natural island amid the cultivated landscape. For a decade, a stainless steel weir installed at the outlet of the gently sloping basin had intercepted only the barest fraction of a millimeter of runoff each year, despite an average annual rainfall of 376 millimeters, most of it falling from May to October during the southern hemisphere winter. Bore holes drilled in the catchment showed that ground water levels weren't rising. Where was the water going?

Different plant types—grasses, shrubs, trees—and even different species among them, vary in the way they capture and reroute the rain that falls on their leaves. Some of the water is intercepted by the canopy and lost as it evaporates back to the atmosphere from wet leaves and bark without ever reaching the ground. Some rain falls between the leaves or drips from them, hitting the ground as "throughfall." The rest is channeled down stems, branches, or trunks as "stemflow," reaching the soil at the base of the bole. Nulsen and his colleagues had set out their funnels to monitor throughfall and ringed the stems with bottles to measure stemflow.

What they saw from their collection bottles was that 15 percent of the rain that falls onto the leaves of a eucalyptus tree simply evaporates away; 60 percent falls through to the ground; and 25 percent runs down the stem. Red dye powder sprinkled on the soil below allowed them to track the fate of that stemflow. After the winter rains, when they excavated around the trees with a backhoe, they found red dye trails penetrating deep into the soil only along channels formed by the tree roots.

These native trees, whose roots extend as deep as twenty-eight meters, were redirecting significant proportions of the rain and storing it around their roots for use during the dry summers. Even the rain that fell on open ground and ran off was slowed by dense layers of leaf litter under trees and shrubs and soaked into the ground within reach of their roots.[7]

The ability to reroute rainfall extends to other woody species in the Australian drylands. Communities of evergreen banksia shrubs and grass trees, which are topped with yucca-like arrays of spiked leaves, can capture and redistribute up to 70 percent of the rainfall that reaches them; and *Acacias* in the arid mulga woodlands channel 40 percent of the rain into stemflow.

One factor, then, that makes the native shrubs and trees of Australia a collective success in dealing with their semi-arid environment is this ability to redirect a good deal of rainfall to their own roots. Another factor is the diversity of their root systems, which allows them to exploit moisture from every level of the soil. Finally, the native plants are mostly perennial and often evergreen. Year-round, they suck in moisture, run their life processes, and transpire water back to the atmosphere from the pores in their leaves. Over eons, this diverse community has evolved mechanisms for capturing, drawing in, and breathing out virtually all the moisture that falls throughout the seasons.

Contrast those talents with the meager abilities of wheat and pasture crops that have replaced native species over most of the landscape. The crops are annuals, taking up and transpiring away water only during the four- to five-month winter growing season. Even at the height of their growth, these uniformly shallow-rooted plants cannot make use of all the rain that falls. Some of the rain pools and runs off the soggy ground, carrying topsoil with it; some filters deep into the plowed earth beyond the reach of crop roots, slowly raising the water table and floating up salt deposits. Ironically, the dearth of rain keeps farmers from planting a second crop during the hot, dry summers, so the occasional summer storms that pelt the stubble fields simply add to the erosion and waterlogging.

This type of land degradation is not limited to Australia. In croplands around the world, waterlogging and salinization of soils are major causes of deterioration. Ironically, however, this damage usually appears on irrigated rather than rain-fed fields. One-third of the world's food is now produced on artificially irrigated lands, and one-third to one-half of those fields suffer moderate to serious salt problems because wasteful amounts of water—far in excess of what the crops can use—are pumped onto poorly drained soils.[8] In southwestern Australia, however, the water supplied to the land did not increase. Instead it was the plants themselves, and with them the architecture of plant canopies, branches, and roots that changed. When they did, even the meager rains of that region became too much for the soil.

At first glance, it seems easy enough to dismiss variations in plant traits as trivial. What farmer watching raindrops beading up on tree leaves or dripping from grass blades could see disaster in the making? How could the rerouting of a bit of rain here and there really have much impact on the water cycle across an entire landscape? Of the total rain that fell over Wheelock's Catchment, for example, only 8 percent was actually captured by trees. Nulsen estimated that clearing the basin for crops and pasture grasses might increase the amount of moisture that percolates beyond the reach of roots by only 4 to 10 percent per year.

Yet after fifteen to twenty years, Nulsen calculated that the excess moisture could be quite enough to raise groundwater and cause salty seepage into low-lying fields. That has certainly proven true in other parts of the wheat belt: Groundwater levels in cultivated fields are as much as seven meters higher than in remnant patches of native vegetation. Nulsen and his team are now watching the scenario play itself out in Wheelock's Catchment. The area was partially cleared in 1985, and despite a drop in rainfall in recent years, runoff through the weir at the bottom of the basin has increased by 40 percent.[9]

In light of such results, the diverse talents of plants are finally getting some of the recognition they deserve. Current recommendations for restoring agricultural lands in the Australian wheat belt to health and sustainable production include replanting buffer belts of thirsty, salt-tolerant native trees and shrubs.[10]

Growth of Deserts

Drylands around the world are also falling prey to another type of degradation: desertification characterized by bush and shrub invasions. (Definitions of desertification vary; some include salinization, although the resulting landscape may look much different than the usual desert of scattered shrubs, cactus, or woody thickets.) Shrubs gain a toehold when human activities damage grasslands or savannas in semi-arid areas. Moving into denuded tracts, shrubs use their own abilities to capture water and rearrange resources, talents that may actually prevent the loss of nutrients and water and sustain productivity and biological activity in these harsh landscapes (see Chapter Seven), but which may also speed the conversion of economically useful rangeland to desert.

Most vulnerable to desertification are the dryland ecosystems that cover about one-third of the earth's land surface, including almost half the African continent. A 1990 study by the United Nations Environment Program concluded that desertification already affects 70 percent of the world's drylands, and much of this degradation can be traced to unsustainable levels of grazing by cattle. Indeed, of the grassy drylands used for livestock range, 73 percent—almost 4 billion hectares—are now moderately to severely degraded. Every year, 9 to 11 million hectares of rangelands finally turn so barren, they are abandoned as useless. Lost productivity on these desertifying lands amounts to $42 billion each year.[11] Finally awakening to the severity of these losses, 102 countries signed a United Nations convention in 1994 to combat desertification.[12]

Combating the transformation of productive land to desert, of course, will require understanding what drives the process in the highly varied regions that are vulnerable. Sadly, there seems to be little hope of reversing the process once it's begun. Along the northern border of North America's vast Chihuahuan Desert, for instance, one area has sunk steadily into decline over the past eighty years even as range managers and scientists have worked feverishly to analyze and halt its deterioration. It was in this most-studied of deserts that researchers began to spot the interplay of human disturbance and plant traits that some scientists believe make desertification a self-reinforcing process.

A century ago, lush expanses of black grama grass carpeted much of West Texas and eastern New Mexico along the northern rim of the Chihuahuan Desert. Since the end of the Pleistocene, no large grazers, not even the native bison, had nibbled the grasses. After the Spanish conquest, cattlemen began to drive herds through these grasslands, but few lingered because they found no water on the plains. When ranchers acquired the machinery in the 1880s to sink deep bore holes for water, however, the attractiveness of the region changed. Soon tens of thousands of cattle were grazing the plains.

Within just a few decades, range managers noticed worrisome changes. Mesquite and creosote bushes were invading the grama grass. Each year the land looked more like the neighboring desert that stretches south into central Mexico. The shrubs had always been around, but their numbers were kept in check by the full carpet of grass, which presented "a very formidable arena for shrub seedlings to get started," notes William Schlesinger of Duke University. The seedlings are no match for the mat of grass roots that captures most of the water and nutrients in the top few centimeters of the soil. Shrub seedlings that do get a toehold seldom survive the periodic wildfires that sweep across unbroken grasslands.

Worried about preserving valuable livestock range, U.S. agricultural officials in 1912 sectioned off more than seventy-eight thousand hectares in the Jornada Basin of New Mexico and set about finding a way to hold back the shrub invasion. They've had little success. Even greatly paring the cattle herds, seeding grass, and battling the shrubs with tractors, drag-chains, fire, and herbicides couldn't halt the conversion to desert, which today is all but complete.

In 1981, under the leadership of ecologists Walter Whitford and Gary Cunningham at New Mexico State University, a group of researchers teamed up to see if they could find out what had tipped the balance in the Jornada, and perhaps in other areas where the desert is advancing at a worrisome pace. Scientists participating in the Jornada Long-Term Ecological Research Project set about learning everything they could about the way this landscape works. As the data piled up, they began to spot a pattern: Beneath the seemingly monotonous landscape, the desert is a heterogeneous place, its soil resources distributed quite unevenly. In 1990, the team published a hypothesis about how desertification starts and why it's hard to stop.[13]

The scenario they describe for the Jornada begins with the massive influx of cattle that followed the drilling of water wells. But similar patterns can be triggered today by off-road vehicles, conversion of dry grasslands to intensive row-cropping, or other human activities that disrupt the grass cover. The key to getting a desert started, they believe, is somehow breaking up the fairly uniform distribution of rainfall and nutrients that grasses enforce. In the Jornada, heavy grazing created patches of bare soil amid the clumps of grama grass. Trampling by cattle, especially around water holes, compacted the soil, making it harder for water to soak in. When the rains came, runoff from these bare and compacted areas carried water, soil, and plant nutrients to cracks and low spots.

Shrubs soon colonized these spots where runoff from intermittent storms accumulated. Their deeper roots allowed them to mine the water and nutrients that had percolated into the ground beyond the reach of grass roots. Once shrubs began to proliferate in this way, their own traits helped drive further changes in the landscape, creating the beginnings of the "islands of fertility" pattern that characterizes deserts. Resources such as nutrients and water that were once distributed among grass clumps no more than centimeters apart now began to get concentrated around shrub-controlled islands one meter across and several meters apart.[14]

Take water, for example. Whether it arrives by throughfall, stemflow, or runoff from barren ground, most of the water that infiltrates the soil in a desert does so under the canopy of the shrubs. Eventually, all the forces of water, wind, and geology seem to conspire to direct nutrients to these shrub islands. Raindrops strike and erode the bare soil between the shrubs. The nitrogen-laden dust that isn't carried away as runoff is swept up by winds, and some of this dust is intercepted by trees and shrubs. With the next rains, the dust washes from the leaves and drips to the ground, where it joins the decomposing leaf litter beneath the canopy to form a nutrient storehouse for the shrub.[15]

Is the conversion to shrub dominance reversible? Schlesinger, now one of the leaders of the Jornada research team, believes that, in practical terms, the answer is no. His team has tried mightily to do it, even to the point of churning and remixing the soil with bulldozers to "rehomogenize" the resources. But that effort and others have failed to turn the area back to grassland. Unfortunately, even if the soil could be made hospitable for grasslands again, half the grassland plant species have already been lost from the Jornada. Thus, there is no easy way to reassemble the pieces to restore this ecosystem to its original functioning, even if ecologists learn how.

Just what causes the boundaries between arid grasslands and shrub deserts to shift in any given region is still a subject of debate. Certainly, climate shifts may be a factor, along with human activities that damage the land. Sadly, climate change projections, based on a doubling of atmospheric carbon dioxide, call for less rainfall over many semi-arid regions, not more, a trend that could accelerate desertification worldwide.

Grass Versus Thickets on the Savanna

Across vast savanna regions of Africa, and to a lesser extent South America and Australia, grasses and shrubs normally maintain a dynamic balance, unlike the Jornada where the two cannot seem to coexist. In fact, a savanna is defined by its mix of woody plants and grasses.[16] Yet these systems, also, can be pushed too far and desertified—converted to thorn bush, cactus, or wooded thickets.

Scientists have yet to find a single, universal mechanism that keeps all savannas mixed. But in some parts of Africa, competition between trees and grasses for water has led to a partitioning of this resource, with each type of plant sending out its roots to exploit moisture from a different level of the soil.

Such is the case across the hot, dry steppes of South Africa's northern Transvaal, where rain falls unpredictably. Some parts of this bushveld host a classic postcard African savanna: flat-topped thorn trees (*Acacia tortilis*) punctuating a vast, open expanse of grass. In other areas, dense stands of broadleaf trees shade nearly one-third of the savanna. Researchers from the University of the Witwatersrand found that the roots of the grasses in both types of savanna communities are concentrated in the topsoil, supporting the idea that grasses get most of their moisture from this upper layer. The roots of the trees mostly plunge beyond those of the grasses, into the deeper soil.

If grasses are truly better equipped to compete for topsoil water, and most trees for deeper water, then the proportion of each plant type in a community should match the rainfall patterns. And it does. The *Acacia* savannas seldom receive enough moisture to wet the subsoil, and as a consequence, grasses cover the majority of the landscape; trees are sparse. On the broadleaf tree savanna, rainfall is heavier, and thus, trees are more abundant amid the grass.

Why is such coexistence possible here but not in the Jornada? Researchers think that because rainfall on the savannas varies greatly from year to year, changing the relative proportions of water in each soil level, neither trees nor grasses have a constant advantage that would allow one to completely eliminate the other. Only when both are in the system does the full water resource get used effectively for plant growth, since grasses cannot reach much of the water freed up by removal of shrubs and vice versa.[17]

Other natural forces help keep the African savanna mixed. Migrating herds of zebra, wildebeest, gazelles, and buffalo have nibbled these grasses through epochal time without destroying them. In balance, both fires and browsing animals such as elephants (a keystone species in the savanna, explored at length in Chapter Seven) have kept the shrubs and trees under control. Until this century, even nomadic herdsmen and their cattle followed the rhythm of the rains across the savanna without destroying the balance of the vegetation.

All that is changing, and now desertification threatens vast tracts of Africa's savannas, too. Booming populations and restricted political borders have forced many nomads into permanent settlements, and their cattle are now grazing and trampling nearby savanna to wasteland. A tourist on safari to Tanzania's Serengeti National Park can easily see desertification in action. Inside the park, *Acacia* savannas stretch to the horizon; just beyond the park borders, bare earth, cactus, and spiny shrubs dominate the land.

Similar patterns are emerging elsewhere in the world. In northern Argentina, just as in the Jornada, the subtropical savannas of the Chaco region have been converted to cactus and thorn scrub since cattle were introduced barely a century ago. In the mulga woodlands of Australia's New South Wales and southern Queensland, formerly grassy stretches have been so destroyed by grazing that even intensive restoration efforts have failed.[18]

Nowhere is the problem so fraught with potential for human suffering as in the savannas of the Sahel. The Sahel shares a six thousand-kilometer border with Africa's Sahara Desert to the north. Over the millennia, the desert margins have advanced and retreated continually with periods of drought and rainfall, and nomadic peoples have traditionally followed the grasses. Today the desert is pushing into the Sahel once again. This advance is more troubling than past shifts because it coincides with a rapid increase of the human population and the growing confinement of nomadic herdsmen and their cattle, sheep, and goats in permanent settlements on the desert fringe. This intensive land use, researchers like Schlesinger fear, could well tip the balance in the drought-stressed Sahel grasslands toward permanent shrub desert.[19] If that happens, a few good rainfall years may not be enough to reverse the desert's advance this time, and that could mean famine and misery on an unprecedented scale.

Water Harvest

In the cases described so far, most water that runs off or percolates beyond the reach of roots is either wasted or troublesome. Yet streams, lakes, and reservoirs would run dry if no water ever escaped capture by plants. As mentioned earlier, one-third of the rain that falls over land does manage to run off or seep through the soils, eventually joining the flows toward the sea. In watersheds where that flow is captured and directed to homes, industries, and farms, it is referred to as water yield. This is water as commodity, water that human societies can put a price on.

The yield depends fundamentally on how much rain and snow falls in a region, and on the seasonal pattern of the rain—whether it falls in sporadic, gentle showers year-round, for instance, or during periodic monsoonal downpours that no plant community can efficiently capture. Thus, although trees are great rain interceptors and transpire a greater volume of water than do grasses, virtually every forest produces streamflow, while undisturbed prairies and savannas do not. Why? Because forests exist in zones of higher rain and snowfall and so cannot always use the vast quantities of water available at any given time.

In any given climate zone and watershed, however, a change in vegetation can dramatically alter water yield. One of the few comparative measurements of water yields from different plant covers in the same basin during the same years began in the 1940s in the San Gabriel Mountains northeast of Los Angeles. During a twelve-year study, watershed managers monitored the volume of water that percolated through soil covered by chamise shrubs, scrub oak, Coulter pines, buckwheat bush, or perennial grasses. During a severe drought, the researchers found that every bit of moisture that reached the soils under the woody plants was used up through evaporation or transpiration. Only the grass-covered plots yielded water. Even in the highest rainfall year, the trees and chaparral shrubs let only 11 percent of the water that entered the soil percolate through, compared to a 40 percent yield from the grasses.[20]

Paying attention only to yield, of course, is grievously shortsighted; another factor to be considered is topography. The results just described encouraged forestry officials in southern California to convert some of their catchments from native chaparral shrubs to grasses. Unfortunately, they picked unstable mountain slopes renowned for their mud slides. The result was dramatic: Triple the soil slippage and erosion rates during severe winter storms where the brush had been converted to grass.[21]

About the same time, another classic long-term experiment was getting started in a very different watershed across the country. When this study began in the mid-1950s, foresters and watershed managers had been engaged in a long-running debate: Will any forest tree work the same on a given site? Are trees completely interchangeable? The questions hinged on whether trees vary in their capacity for water interception and transpiration.

To get answers, loggers clear-cut mature hardwood forests of native oak and hickory in two watersheds in the southern Appalachian Mountains of North Carolina, then replanted them with white pine seedlings. By 1973, when the pines had filled out their canopies, those two watersheds were yielding 20 percent less streamflow than equivalent catchments still forested by hardwoods. What Wayne Swank and James Douglass of the U.S. Forest Service Coweeta Hydrologic Laboratory near Franklin, North Carolina, found was that transpiration and evaporative losses from pines are greater because this species retains most of its needles year-round. In contrast, the dormant hardwoods stand leafless through the fall and winter, and their bare trunks and branches allow more rain to reach the soil and seep to the streams.

The study was not meant as a purely academic exercise but as a guide to watershed management decisions. Swank and Douglass's results showed that the conversion of just sixteen hectares of forest from oak-hickory to pines cost 23 million liters in lost water in a single year. The amounts would be different in other climates, yet a conversion of any deciduous forest to conifers is likely to result in lower streamflow.[22]

The finding is sobering in light of the fact that large swaths of the world's temperate forests are being converted from diverse broad-leafed woodlands to pine and fir plantations. Indeed, the interests of water supply managers and timber producers, as well as conservationists, fishermen, hikers, and others concerned about the fate of forested basins, don't always harmonize, and watersheds are often patchworks under multiple ownership. Timber companies prefer conifers because these fast-growing trees allow them to harvest a greater volume of wood from the same nutrient supplies over a shorter time period. But there are also circumstances when water districts have reason to prefer conifers. Why? Although they reduce overall water yield, they also reduce peak stormflows and so lessen the potential for flooding. [23,24]

Cloud forests, which are dense tracts of evergreens rising thousands of meters above sea level, contribute to yield, but in a novel manner: they literally comb water from the clouds. In the Cascade Mountains of Oregon and other coastal slopes or mountain summits, for instance, firs are often shrouded in swirls of fog or immersed in wind-driven clouds. Water condensing on the needles and dripping to the ground can increase precipitation enormously. From the mountainous forests of

Hawaii to the Bavarian Alps, trees can sometimes scavenge more moisture from the clouds than the clouds yield directly as rainfall. Cutting such forests not only decreases streamflow but often changes the functioning and character of the landscape.[25,26]

Such is the case along the southeast coast of Mexico, north of Veracruz, where damp winds from the Gulf of Mexico sweep inland and rise up the slopes of the Sierra Madre Oriental range. The warm winds once drenched and nurtured thick rain forests on the eastern slopes, then rose toward the summit, cooling and forming dense fogs. Lush forests on the high plateaus intercepted their water from this fog. Farmers crowding into this region in search of land, however, cleared the plateaus years ago to plant maize, then moved down the eastern slopes, cutting and burning the forests to make new fields. As luck would have it, their dreams of well-watered crops failed. The water simply vanished with the trees, and now the slopes support only cactus.

When Yale University botanist Hubert Vogelmann studied the area, he found that the gulf winds still drive moist fogs across the slopes, even in the dry season. Without trees to intercept the fog drops, however, this moisture simply evaporates as it sweeps inland over the now-arid ground.[27]

Alien Trees, Water Yield, and Erosion

Throughout much of history, humans have behaved as though plants were virtually interchangeable on the landscape, whether the plants were different tree species, such as pines and hickories, or completely separate life forms, such as eucalyptus trees and wheat. Yet plant species do not all function in the same ways. Even plants moved between similar climate zones may function in ways that devastate ecosystem services in a new setting. For example, thirsty trees like Australian hakeas and *Acacias* do a good job of keeping groundwater levels low at home, but transplant them to the climatically similar mountain watersheds of southern Africa and those traits can be costly, both to the native ecosystem and to the citizenry.

European settlers to South Africa liberally supplemented the diverse but treeless native scrub, or fynbos, of the Cape region with imported trees and shrubs for timber and horticultural uses. Trees from similar Mediterranean climates—pines from southern Europe and hakeas and *Acacias* from Australia—that were thought to be well-suited to South

Africa's climate have done so well that they have spread across the region, forming dense stands that crowd out the shorter native shrubs, such as proteas. Such tall trees obviously require significantly more water than do shrubs. The result? A precipitous drop in water yields by as much as 30 percent.

Controlling the spread of these alien species is a particularly vexing and costly problem for watershed managers. Fires, which are periodically set to keep fuel loads down in the native fynbos, actually further the spread of the invaders, which release their seeds in the aftermath of fire. On the steep slopes of Table Mountain, which rises from the middle of urban Cape Town, exotic pine forests long ago displaced much of the fynbos. Today, people enjoy the forests, and a vocal part of the citizenry strongly opposes their removal. One consequence of the forests remaining is that when wildfires rage, they do so with much fiercer intensity than if the slopes were shrub-covered. Without fast-recovering shrubs to protect the scorched earth, severe erosion follows, with the city spending inordinate amounts of money after each fire scraping sediment from streets, storm drains, and other urban structures.[28]

There seems no limit to the insidious ways that plants, moved to inappropriate settings, can alter water cycling. Some deep-rooted invaders that colonize watersheds and river banks act much like dams, trapping enough sediment to choke a river channel. The banks of the Colorado River bear witness to this type of invasion.

Photos taken in the Grand Canyon in 1871 by explorer John Wesley Powell's expedition show the rocky banks of the Colorado River largely bare of trees. However, at about the same time, an Asian shrub called tamarisk, or saltcedar, was being introduced into the United States. Barely a century later, saltcedar covered an estimated half-million hectares throughout all the major river drainages of the American Southwest, from the Colorado to the Green, Gila, Virgin, and Rio Grande. By comparing current aerial photos with historical photos and surveys, geographer William Graf of Arizona State University calculated that tamarisk had invaded along the upper reaches of the Colorado and Green Rivers at the rate of about twenty kilometers a year starting in the 1920s. The result: The river channels have been narrowed by an average of 27 percent.

The deep-rooted shrubs invade banks and sandbars that once would have been colonized by native salt grass, dwarf willows, and cottonwoods. In the past, the grasses and willows would have been swept downstream, along with the sandbars, in periodic floods. But the well-anchored saltcedars hold their ground—especially now, in the reduced flows of these heavily dammed rivers—they catch even more sediment, and build the bars higher and wider. The result is more frequent flooding since even modest flows now spill out of the narrowed channels.[29]

Erosion and Nutrient Runoff

Despite very real hazards from chemical pollutants, the runoff of silt- and nutrient-laden water is by far the leading cause of degradation in U.S. rivers and lakes today, and similar problems are occurring worldwide.[30] Usually the culprit is poor agricultural practices that strip vegetation from the land. When rivers run gray with mud or lakes are befouled with pond scum, it often means someone upstream has ravaged forests, packed too many cattle onto the range, or cleared sloping land for crops.

The world's farmers lose 24 billion tons of topsoil each year.[31] Yet many countries don't have the money or expertise to monitor soil losses on lands where they occur, since the loss of 6 tons represents a barely perceptible one millimeter of topsoil from a hectare of land. Instead, the telltale signs of erosion show up in rivers. Erosional losses and flooding are particularly severe in China and India. The Yangtze River carries away 506 million tons of sediment a year, 28 tons for every hectare of cropland, cities, villages, or wildland it drains. China's Huang He (Yellow) River and the Indus River in Pakistan each carry off 25 tons per hectare of basin, while the Ganges hauls off 11 tons per hectare.[32]

In Africa, food production has been faltering at a worrisome rate. A United Nations Food and Agriculture Organization study of farms in Zimbabwe found that even well-managed maize fields lost 10 tons of soil per year. In village plots the losses rose to 40 tons, and degraded rangelands surrendered 100 tons of their topsoil per hectare to winds and water each year.[33]

It would be wishful thinking to dismiss erosion as a problem of peasant farmers trying to coax food from marginal lands. As a result of U.S. government incentives instituted in the 1960s and 1970s to increase commodity crop production, soil losses on U.S. farms averaged 18 tons per hectare by the early 1980s. Invisible in that average are some staggering losses. In the dryland region along the southern Washington and

Idaho borders, beans, lentils, barley, and wheat are planted on hillsides with gradients from 15 to 25 percent. Not surprisingly, soil losses from these fields are high—113 to 227 tons per hectare. Even in the midwestern corn belt, where land is flat, soil losses run as high as 45 tons per hectare per year.[34]

In 1985 the U.S. Congress created a Conservation Reserve Program designed to cut these losses by paying farmers to put highly erodible land into less intensive cropping, or plant it with trees or grass. While the program is considered a success, few nations have the luxury of paying farmers not to use marginal land. In fact, in many parts of the world, farmers see little choice but to plant on hillsides. They chop and burn the natural forests, plant maize crops for a season or two, then seed the ground to pasture grasses after the soil deteriorates. In the state of Jalisco on the Pacific coast of Mexico, for instance, farmers clear plots on slopes that average twenty-two degrees, despite soil losses that range as high as 130 tons per hectare. Researchers hoping to control this erosion and help farmers get longer-term use of the land tried adding mulch from nearby forests to some experimental maize plots and planting buffer strips of grass below other fields. The mulch worked best, cutting erosion by 90 percent, helping to retain more water on the plots, and increasing crop production by 30 percent. Yet soil losses still ran nearly 6 tons per hectare, an unsustainable level for tropical soils.[35]

The results clearly show that neither maize nor grasses can duplicate the services of the original tropical forest. With its two hundred plant species, the forest lets virtually none of its soil or water escape and sustains a lush productivity year-round. Needless to say, the utter destruction of the forest for a few years' worth of unsustainable grain crops is a tragic trade.

Sediment flows, of course, are only half the runoff problem created by modern agriculture. Since the mid-twentieth century, chemical fertilizers have powered an enormous increase in world food production and are now intricately linked to agriculture. Between 1950 and 1989, for example, the world's farmers increased their annual fertilizer use from 14 million tons to 143 million tons.[36] Too much of this fertilizer, however, never contributes to crop growth. Instead, it takes a terrible ecological toll.

The toll exacted by these fertilizers occurs off-site and thus is largely invisible to farmers, often showing up far downstream in rivers, lakes, and coastal waters. One of the major fertilizer ingredients, phosphorus, binds to soil particles and is carried away with surface runoff. Another, nitrate, is highly water soluble and leaches from nitrogen fertilizers, animal wastes, and organic matter, seeping through the soil into streams or groundwater. (Nitrates are major plant nutrients in the soil but serious pollutants in drinking water supplies. They have been correlated with increases in lymphatic cancer, and, at high levels in drinking water, nitrates can cause a type of anemia in infants.)

When excessive amounts of these nutrients pour into rivers, lakes, or wetlands they create a condition called eutrophication, fertilizing the growth of algae just as they would have done for crops or trees had they been held on the land. Blue-green algae, or pond scum, is a particularly visible sign of eutrophication. Some species of these algae can be toxic to people or livestock. Others just imbue the water with an obnoxious frog-pond taste. If the algae bloom too profusely, tiny animals (zooplankton) that graze on the algae cannot keep algae populations in check; the algae die and sink to the bottom, where they form deep layers of muck. Bacteria then move in, and, working furiously to decompose the dead matter, use up the oxygen in the bottom waters. Loss of oxygen then renders the lake inhospitable to most oxygen-dependent organisms, including fish and invertebrates. If the eutrophication is severe, the lake eventually may be declared dead.

Fortunately, unlike many ecological disruptions, this type of degradation is reversible if the nutrient runoff can be halted. One key study on blocking nutrient runoff took place near Chesapeake Bay on the U.S. Atlantic coast, a region where farmers spread 700 million pounds of commercial fertilizers on their fields each year.[37] William Peterjohn and David Correll of the Smithsonian Environmental Research Center in Edgewater, Maryland, monitored the chemistry of surface runoff and shallow groundwater seepage from a cornfield along the western shore of the bay. Like most croplands, the cornfield was quite "leaky," retaining little of the nitrogen fertilizer it received. Most of the nutrients not absorbed by the crop and harvested as corn were lost in surface runoff. Yet when that runoff reached a small belt of hedgerows and deciduous forest nearby, the gauntlet of sweet gum, red maples, and understory shrubs stripped most of the nitrogen and phosphorus from the water in the first nineteen meters of its passage. Groundwater flows through the root zone of the forest also resulted in a 50 percent drop in dissolved nitrate reaching the river.

The amount of material removed by this one narrow band of forest led the researchers to suggest "coupling natural systems and managed habitats within a watershed" to buffer open waters from pollution.[38] This is a key service of biodiversity that, until recent times, could easily be taken for granted. Most rivers are bordered naturally by some type of riparian or floodplain forests that buffer aquatic habitats from activities on land. Yet, in heavily populated regions throughout the world, land clearing has eliminated most of these. Now, people run cultivated fields, golf courses, lawns, buildings, parking lots, and pastures right to the water's edge, and pollutants seep or trickle into waterways across the length of the landscape. Clearly, restoring forested corridors could do much to hold nutrients and sediments on the land and improve the quality of lakes, rivers, and wetlands.

Consequences of Overfishing

Because predatory fish eat smaller fish and don't graze on algae, no one suspected until the early 1980s that these carnivorous species had a major influence on the water quality of lakes. However, researchers looking at how algal productivity varied from one lake to another around the world came upon a puzzle. Often lakes with the same amount of nutrients pouring in had vastly different rates of algal growth. Some other factor seemed to be exerting equally powerful control, causing lakes with similar supplies of nitrogen and phosphorus to vary widely in water quality.

More than a decade ago, Steve Carpenter of the University of Wisconsin and his colleagues proposed that the key could be found in the food webs of these lakes. Large populations of predatory fish, such as pike, bass, walleye, and muskellunge in a lake hold down numbers of smaller prey fish such as minnows, which in turn keep algae-eating zooplankton in check. Algal growth, finally, depends on nutrient supplies. When pike are abundant and minnow numbers are reduced, the researchers reasoned, zooplankton thrive and keep the standing stocks of algae lower than the maximum allowed by nutrient levels. The researchers proposed that a change in just one of these links in the food web—for instance, heavy fishing that reduces pike numbers—sends a ripple effect through the system that eventually affects the amount of algae in the lake.

Carpenter and James Kitchell, also of Wisconsin, set up a test of this ripple effect, called a trophic cascade, in a series of lakes in Wisconsin. The most dramatic shift came in Tuesday Lake, which had no bass but had abundant minnows. The researchers removed 90 percent of the minnows by weight and introduced stocks of largemouth bass from a nearby lake. The abrupt decline in minnows, which had kept the zooplankton numbers low, caused populations of these tiny grazers to expand dramatically and knock down the high densities of algae by 80 to 90 percent.[39,40]

Restoring predatory fish has already been used successfully to control water quality problems in some lakes and reservoirs. In terms of money, practicality, and sheer aesthetics, putting fish into a sick lake easily wins out over engineering alternatives, which include dumping aluminum sulfate into a lake to bind phosphorus and carry it to the bottom or placing mechanical aerators on the bottom to reoxygenate the waters. In a few cases, polluted lakes have even been handled literally like sewage, their waters pumped out, cycled through on-shore waste treatment plants and then poured back.

Fish stocking is no cure-all, though, especially without improvements in land management. No fish in Wisconsin could be expected to counter the nutrient runoff from the manure of a heavily stocked dairy farm. Carpenter's team believes cleaning up degraded lakes requires land use changes that reduce such runoff, as well as a series of biological fixes. These include creating vegetated greenbelts about fifty meters wide around the shoreline; restoring shallows and wetlands needed for fish breeding and nursery areas; and, of course, restoring and maintaining stable populations of native big game fish.

This last fix may require some sacrifices on the part of fishermen. Tight limits on fish catches, or even catch-and-release requirements, are at first a hard sell to avid anglers. "Some of our restorations have actually been derailed by fishing pressure," Carpenter admits. Yet local fishing clubs and other people who use a lake regularly are often supportive once they understand the goals of the regulations, he says. Conserving fish means a healthier lake and more fish for the long term. Tourist-oriented businesses and anglers who visit the lakes only on holiday are a harder sell. Carpenter points out that adding a touch of sociology in the management plan helps. It's easier to create a catch-and-release ethic for the big, long-lived fighting fish like muskellunge and Northern pike and save them for cleanup duty, for instance, if fishermen are allowed to keep tastier catches, such as walleye, or even encouraged to harvest zooplankton-eating panfish like yellow perch and bluegills.

Alien Fish and a Dying Lake

In a few cases, introducing an exotic fish species to a fresh water lake has led to steep declines in water quality. The classic example occurred in East Africa with the introduction of predatory Nile perch to Lake Victoria, a tropical lake with a much different food web than most temperate lakes.

Lake Victoria is the largest body of fresh water in Africa; worldwide, it ranks second only to North America's Lake Superior in size. Yet human activities have brought catastrophic changes to the lake only a century after Europeans colonized the region. At first, most of the changes resulted directly from land conversions and overexploitation of resources. Large tracts of watershed forest in Kenya and Uganda, for instance, were cleared and planted in tobacco, sugar, cotton, tea, and coffee, resulting in serious runoff to the lake. Wastes from sugar factories and other industries were also discharged into Lake Victoria, along with sewage from a burgeoning human population.

By the 1950s, overfishing had already eliminated several popular fish species. Cores of lake bottom sediments taken in 1990 by members of the Lake Victoria Research Team indicated that, largely unnoticed, water quality and algal populations had begun to shift by the 1920s. But it was a bit of human manipulation that indirectly pushed the lake over the edge.

The native fish community of Lake Victoria included no top predators. What it did have was four hundred species of little cichlid fishes, five to ten centimeters long, that provided abundant protein to millions of people in the three nations around the lake and also served as a living laboratory for evolutionary biologists. Among this profusion of closely related species were specialists in every feeding strategy, from crushing snails to eating crabs, insects, eggs, or baby fish. Most important, however, the vast majority of the cichlids grazed on algae or nibbled detritus at the lake bottom, thus helping to prevent algal blooms even in the face of increasing nutrient loads.

British colonial officials in the 1940s and 1950s debated the idea of introducing a large carnivorous fish, such as the Nile perch, to feed on the cichlids, a move that would create a sport fishery as well as provide a better table fish for European tastes. Before the debate was settled, Nile perch, some nearly two meters long, began showing up in fishermen's nets. Once perch had been clandestinely introduced, fisheries officials actively stocked them. For twenty years, however, the perch remained a minor presence in the lake. Then, in about 1980, for unknown reasons, perch populations exploded. The number of cichlids fell sharply, and within a decade, at least half the original species were believed extinct.

Unfortunately, many of those lost cichlids were algae grazers; in their absence, algal populations exploded, becoming five to ten times more abundant than when the perch were introduced. The result is a severely eutrophied lake, with badly oxygen-depleted bottom waters. Now seasonal turnovers of warm surface waters and cooler bottom waters often trap fish in anoxic zones and cause major fish kills.

The radical change in fish populations and loss of biodiversity have also greatly altered the nutrition of people around the lake, and indeed, the very structure of village life. Traditional fishermen poling wooden canoes cannot follow the commercial trawlers onto the open lake where the Nile perch abound, and their nets are not strong enough to snare perch. Their catch of cichlids and other small fish in the bays and lagoons is meager, and officials have begun to worry about protein malnutrition among villagers, who cannot afford to compete with the export market for a share of the 200,000 to 300,000 tons of perch filets shipped out of the region each year. Men can leave their canoes and find work in the commercial fish processing plants that have mushroomed around the lake. Women, who traditionally processed and sold the village fishermen's catch, are reduced to purchasing perch carcasses from the filleting plants to fry and sell along the roads outside the larger towns. These scraps are the only fish most lakeside people can now afford.

International teams of scientists as well as aid donors and government ministers from all three lakefront nations—Kenya, Uganda, and Tanzania—have huddled for years discussing Lake Victoria's problems. All worry that the degraded lake won't be able to support even the commercially valuable Nile perch population much longer. Improving the health of this lake is thus a matter of life and death for many villagers. The strategies most often talked about for salvaging Lake Victoria sound much like those in Wisconsin: improving land management to reduce nutrient runoff; reforesting the shorelines; rejuvenating the papyrus swamps and wetlands; and finding some way to regain the services of fish, in this case, grazing fish. Only a few small-scale demonstration projects have been implemented, and many ecologists are pessimistic about restoring this ecosystem to healthy functioning.[41,42]

Beavers, Fallen Logs, and Healthy Streams

While lakes are particularly vulnerable because materials that enter them are often flushed downstream slowly, if at all, the opposite holds for streams and rivers. Their health and liveliness often requires slowing down the flow, creating a variety of habitats both within the stream and along its floodplains and assuring that materials don't leave the system too quickly. The organisms that do the best job of this are beavers and trees.

For extensive stream remodeling that creates new wetlands, waterfowl habitat, fish spawning pools, and settling ponds for sediment, nothing matches the work of beavers. In a world where fully half of our original wetlands have been drained, filled, or degraded, the creative powers of beavers should be more celebrated.

Beavers and their rodent relatives, coypu, once lived along streams throughout the temperate and boreal forests of Europe, Asia, and both Americas. Various estimates put their densities in North America before the arrival of European settlers at 60 to 400 million animals. From the arid regions of northern Mexico to the Arctic tundra, beavers thrived wherever there was water. However, fur trappers and landowners have vastly reduced the numbers of beavers in most places. In many areas, the loss has caused dramatic changes in the riverine landscape. In eastern Oregon, for instance, when untended beaver dams finally failed, faster stream flows cut narrow channels and lowered the water table, eliminating most of the deciduous forests along the stream banks.[43]

Beavers have been reintroduced across the United States, Europe, and Russia in recent years, and in areas where they have been allowed to reestablish in force, it's still possible to watch how quickly they can transform a river system.

One of those places is Minnesota's forested Kabetogama Peninsula in Voyageurs National Park, where beaver numbers have been increasing since the 1920s. Aerial photographs taken over a sixty-three-year period have allowed Carol Johnston and John Pastor of the University of Minnesota and Robert Naiman of the University of Washington to track the

Beavers can transform a forest into a diverse mosaic of meadows, wetlands, ponds, and woods that influences the landscape for centuries.

animals' expanding influence on nearly three hundred square kilometers of stream-laced boreal forest. Over the years, beavers cut down willows and aspens to construct more dams, forming ponds that each inundated from one to three hectares along the stream. The animals also thinned the forests one hundred meters out from their ponds, each one cutting as much as a ton of wood each year to be carried back to its lodge. As deciduous trees in the flooded areas rotted to snags, great blue herons claimed them for rookeries. Ospreys came to fish in the new ponds. As the open waters increased, wood ducks, mallards, and black ducks landed to feed, and white-tailed deer and browsing moose frequented the clearings. Sediment and nutrients brought down by the streams settled out in the ponds, and rotting vegetation formed a rich reserve of organic matter. As older dams broke, some ponds drained, exposing a rich staging ground where meadow plants quickly colonized. Other old ponds became backwaters, premier fish spawning sites.

Within six decades, the number of beaver ponds had grown from 64 to 835, and the creatures had converted 13 percent of the peninsula from a uniform stand of forest to a diverse mosaic of habitats in a woodland setting: moist meadows, wetlands, and ponds. The effect of their work is likely to remain visible on the land and its drainage networks for centuries.[44-46]

Like beavers, trees deserve respect for their work in shaping the riparian landscape. At one time, most rivers wandered erratically through their floodplains, flowing in wide, braided systems of sloughs, islands, and multiple, shifting channels. Lush forests grew along their banks, and trees falling into the channels frequently narrowed or diverted the flow, creating new meanders. Humans, however, want rivers to flow in single deep channels that can be navigated by ships, bridged by highways, banked and dammed for flood control, and demarcated on maps as dependable boundaries between properties or territories.

Consequently, most major rivers have been "cleaned up" and narrowed substantially by dams; channelization; and removal of fallen logs, snags, and much of the standing forest. Along the lower Mississippi, a river immortalized by Mark Twain for its treacherous shoals and meanders, James Sedell of Oregon State University found that more than eight hundred thousand snags had been cleared in a single fifty-year period. These were not small logs: Most of the snags were cottonwoods and sycamore trees averaging 35 meters long and 1.7 meters across at their base when they were dragged from the channel.[47]

Although humans have usurped the power that trees once wielded over major rivers, woody debris still plays a key role in the structure of headwater creeks and streams. Just as important, trees and their litter supply much—sometimes all—of the organic matter that supports the food chain in these waters, from insects to salmon and trout. Without fallen trees and log or beaver dams to slow the current, this litter of leaves, needles, twigs, cones, branches, and bark would be of little food value. Microorganisms from fungi to insects would not have time to decompose the material were it not allowed to accumulate behind obstacles.

Moreover, log dams slow and divert the flow, helping to form riffles, eddies, waterfalls, and protected pools where salmon spawn and trout seek refuge from the summer heat. When water is slowed by log dams, sediment settles out, just as it does in beaver ponds, keeping rocks and gravel in much of the streambed free of sediment.

These gravel beds provide attachment sites for eggs and nymphs of insects. Fly fishermen know that trout lurk in the rocks, snapping up hapless nymphs that lose their grip on the gravel and drift downstream on the current. In spring, as the nymphs emerge as winged adults, trout rise from their pools to snap up successive hatches of may flies, salmon flies, caddis flies, stone flies—and the imitations lofted to them by fly fishermen.[48]

How long the debris dams that support this food web endure depends in part on the types of trees that line the stream. Hardwood logs will rot in just a few years. Conifers, because they contain more lignins (complex polymers that make wood rigid) and other hard-to-decompose compounds, may last a century and provide great stability for the stream community.

Downstream in larger rivers, tree canopies don't reach far over the channel, and the sparse litter that drops into the swift current seldom comes to rest in backwaters, where it can be trapped long enough to decay. Instead, river dwellers from invertebrates to fish must rely for much of their sustenance on the work of upstream creatures that break down trapped leaves and wood and release fine organic particles and dissolved materials into the flow.[49,50]

How much does logging along the banks hurt these upstream communities? Silt running off cleared land certainly smothers gravel beds and muddies the waters. When researcher Sedell resurveyed watersheds that had been logged fifty years earlier, he came upon significantly fewer pools per mile of stream. There are probably changes in fish populations in those pools, too; certainly there are decreased salmon numbers. However, increased fishing pressure, introduced predators, and the massive damage inflicted on sea-run populations of salmon and trout by dams and reservoirs make it hard to tease out the effects of forest losses alone. Fishermen who find no chinook or steelhead in their favorite mountain streams have a hard time apportioning the blame.

Mangrove Forests and Coastal Fisheries

In the tropics as well as in temperate streams, overhanging vegetation may supply the bulk of the organic matter that fuels the aquatic communities below. The blackwater rivers of Amazonia, for instance, are both acidified and tinted dark by organic substances leached from decomposing leaf litter and from the peat-bog soils of annually flooded forests. The acidified water discourages the growth of aquatic plants like algae or water hyacinths, and a traveler will find these sections of the

Amazon blessedly mosquito-free. The few fish species that make a living in blackwater rely on beetles, fruit, seeds, and other materials that happen to tumble into the water from the tropical forest overhead. When these forests are cut, the aquatic community's lifeline is also severed, and little can survive in the rivers.[51]

Downstream in the tropics, along river deltas and lagoons, flooded mangrove forests perform many of the same roles that trees along temperate streams do: sopping up nutrients and sediments that run off the land, stabilizing shorelines, creating habitats that nurture aquatic organisms, and sometimes supplying much of the organic matter that supports the food chain in nearby coastal waters.

Mangrove ecosystems are themselves a diverse mosaic of species and forest types, each preferring a different frequency of immersion by the tides. In the New World tropics, the two most extensive mangrove forests are fringe and basin. Fringe forests dominated by red mangroves grow closest to the shoreline, where they are routinely flooded at high tide. Further inland lie the basin forests dominated by black mangroves, where standing water and organic material are flushed only sporadically into adjacent estuaries. A diverse array of marine animals, including crabs, fish, and shrimp, find sustenance and refuge at various life stages amid the distinctive stilt-like prop roots of the red mangroves; above them in the canopies, herons, egrets, and spoonbills create rookeries.

Most of the mineral nutrients, such as nitrogen and phosphorus, as well as sediments that arrive from upstream, are trapped by the mangroves. This prevents an outwash of materials that could smother or eutrophy shallow coastal waters. Yet the forest's own leaf litter and debris, washed out daily on the tide, subsidizes the production of fish and shrimp offshore.

The importance of mangrove leaves and, more controversially, dissolved organic matter from the basin forests has been studied intensely for more than two decades. One thing that is not in question is the importance of red mangroves in supporting the sport and commercial fisheries of south Florida. Classic studies reported twenty-five years ago showed that fish in those waters were feasting on leaves washed out on the tides; cut open a fish and you could find leaves in its innards. Obviously, the red mangroves are the most prolific and visible exporters of bulk carbon, shipping out large quantities of whole leaves, twigs, and other material daily on the receding tides. The findings have won some legal protections for red mangroves in the United States.

Black mangroves, however, weren't seen as a significant source of fish food because their waters appeared to connect with the sea only a few times a month. Then, beginning a decade ago, Robert Twilley of the University of Southwestern Louisiana and other mangrove researchers braved the heat and mosquitoes of September, which few had apparently done before, to find that during that month, tides actually flush the basin forests daily, sending plumes of dark, tea-colored water into the estuaries. Since black mangroves cover two to three times as much of the landscape as red mangroves, theirs may be the largest contribution of carbon to the local food web. But this organic soup can't be tracked visually on its path through the food chain as leaves can, so some scientists still question its importance.[52,53]

The richness and diversity of mangrove species in the Old World tropics exceeds that of the Americas, yet the role of mangrove debris as a food source for fish in those regions is even less clear than in the New World. Knowing the precise functional worth of these complex forest systems would certainly help counter the destructive exploitation now taking place. Unfortunately, mangrove forests occupy prime coastal land, which is being cleared at an alarming pace, from Florida to Ecuador to the Philippines, for waterfront housing, tourist attractions, shrimp and fish farms, rice cultivation, timber, charcoal production, tannin extraction, and numerous other uses. Losses so far include 80 percent of the mangroves in the Philippines, 50 percent in Thailand and Indonesia, and 32 percent in Malaysia.[54]

"When you start dealing with economists, you have to take carbon numbers and translate them to dollar bills," Twilley notes. "The question becomes, if we cut 10,000 hectares of mangroves, how many pounds of fish are we going to lose? We can't answer that yet." We also cannot quantify the damage that might be caused to coral reefs offshore as they slowly smother under loads of sediment and nutrients that the mangroves once intercepted, or to coastal land as it is eroded.

One of the most ironic and short-sighted reasons for the clearing of mangroves is to create aquiculture facilities, especially shrimp ponds. Ecuador had 40,000 hectares of shrimp ponds in 1983, most of them built inland behind the mangroves. In less than a decade, that figure tripled. The export value of the 1991 shrimp crop was estimated to be $482 million, second only to oil as an export commodity in Ecuador. Most of the new facilities were built in intertidal areas cleared of mangroves, as close to the shore as possible so that lagoon water can be pumped into the pools and concentrated wastes flushed back into the estuary.[55]

Yet the mangroves being cut are those that must protect the estuaries and coastal waters from nutrient loads like shrimp-pond wastes. The question is: At what point will the system break down, causing a snowballing decline in water quality, fisheries, and offshore coral reefs? Like so many other aspects of the relations between biological losses and the water cycle, the question remains unanswered. The answer, when it does come, may come hard.

CHAPTER FIVE

The Vitality of the Soil

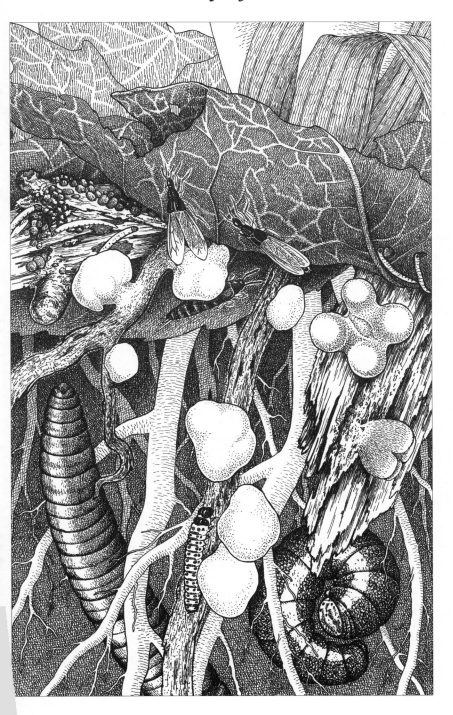

Overleaf
Scientists still know little about how the myriad unseen and largely unknown species in the soil contribute to its vitality.

*O*n Washington's Olympic Peninsula, most people's eyes have been focused on a long-running battle over the fate of centuries-old forests. A struggle less noticed has been taking place on two of the many clearcuts that scar the region. Douglas fir seedlings replanted in one of these clearcuts already tower over most of the researchers who monitor their progress. On a clearing across the road, only a few scraggly, meter-high trees dot the expanse of weeds and stumps. Most people driving past would simply assume that the healthy young trees had a five- to ten-year head start, while the other clearing was recently logged. Yet the two plots were logged and replanted only a year apart—in 1987 and 1988. One plot represents a forest comeback; the other a struggling failure.

In laboratory culture dishes and under microscopes, the evidence clearly shows that loggers unwittingly destroyed something more than trees on the second plot. Beneath the majestic Douglas firs, beneath the dense understory of salal shrubs, sword ferns, and twinflowers, this forest had been nourished by an underground community hundreds of times more biologically diverse than the lush aboveground community once visible to us. After loggers cut and dragged away the firs, ripping the understory with their skids, they left behind a subsoil landscape as dramatically changed as the visible one.

Without the tiny organisms that dwell beneath the surface, this patch of soil has forgotten how to make a forest.

The earth may be dominated by water, but its name refers to the soil, to that thin layer of life-sustaining "earth" that living things helped to create and now must work ceaselessly to maintain atop the rocky bulk of the planet. It is soil that lies at the heart of the ancient cycle of dust to dust, that makes it possible for raw elements garnered from air, water, and bedrock to be assembled, broken apart, and reassembled into generation after generation of fir trees, ferns, microbes, and humans. The soil is a complex and richly diverse ecosystem, yet it is too often treated as dirt—an inert substance to be poisoned, plowed, scraped bare, and still coerced into supporting any plant sown into it. The more scientists learn about soil, the more they realize that maintaining the vitality and character of the community living within it is crucial to what grows on top.

Soil has been called "the biologically excited layer of the earth's crust."[1] It is at once organic and inorganic, jointly created by and composed of both physical elements and living organisms. Soil varies greatly from place to place depending on the climate, the underlying rock, and the organisms present on the land. Around the world, ten major orders of soils containing perhaps one hundred thousand variations have been identified and parsed into a taxonomic system equivalent to the one used to classify living organisms. Although the composition of soils varies widely, roughly 45 percent of an average soil consists of minerals. Most of these are mined from bedrock by wind, water, ice, tree roots, and lichens, which cleave rock into pebbles, then crumble it into grains. Organic acids exuded by roots and microorganisms then decompose these grains into mineral elements, including essential plant nutrients such as phosphorus, potassium, calcium, and iron.

Water occupies another 25 percent of the soil volume, and air 25 percent. Only a relatively small proportion, roughly 5 percent, is composed of organic matter, living and dead. Despite its relatively small proportion in most soils, any gardener knows that the quality and quantity of this organic matter is the key to the structure and fertility of the soil. The dark and caky humus, which is a partially rotted compost of leaves, twigs, fruit, animal carcasses, feces, and other organic matter in the process of being broken down by organisms ranging from earthworms to bacteria, supplies the carbon and nitrogen reserves that fuel and nourish both soil organisms and aboveground plants and animals. Moreover, the humus provides vital structure to the soil, helping to "glue" soil particles into spongy crumbs and creating pores that hold air or water. It is this structure that makes the soil a habitat for bacteria, fungi, and tiny soil animals. Soil depleted of organic matter or compacted by machinery or the hooves of livestock becomes less permeable to water and air and thus less able to support either plants or oxygen-dependent microbes.[2]

Physical forces such as erosion, weathering, and leaching take a constant toll on the soil, especially—as the previous chapter described—when humans strip away natural vegetation during ill-considered agricultural ventures. A recent study sponsored by the United Nations

Environment Program found that, since 1945, soils across 17 percent (2 billion hectares) of the planet's vegetated land have been degraded by human activity. About 300 million hectares of that is so severely degraded that it is considered useless for agriculture unless "major international assistance" can be provided to reclaim it.[3]

Because of this ongoing toll, both natural and human-caused, the creation and maintenance of fertile soil is one of the most vital ecological services the living world performs. However, creating a few centimeters of new topsoil from disintegrating bedrock and organic decay can take several hundred to one thousand years, meaning that lost soil is not a renewable resource on any time scale practical to today's farmers, or even their great grandchildren. Thus, the most critical ecological processes are those that renew and maintain the life-supporting functions of the soil.

For instance, the mineral and organic contents of soil must be replenished constantly, as plants consume soil elements and pass them up the food chain. This process of replenishment reflects a subtle synchrony between the diversity of life above and below the ground. Plants are not just takers but givers, too. They not only shelter the soil from the harsh actions of wind, sun, and rain, but also supply it with litter and detritus, thus replenishing its carbon and nutrient stocks. In addition, anywhere from 10 to more than 50 percent of the carbon that plants capture through photosynthesis eventually passes from roots and symbiotic root fungi into the soil, where it feeds the microbes that convert the plants' detritus back into reusable nutrients.[4] (Of course, this bounty can also feed parasites and pathogens that are part of the soil community.)

In any given climate, on any type of parent rock, it is the interactions between the soil and plant and animal communities that create and maintain the characteristic soils of a region, from the dark, rich organic soils of grasslands to the acidic, infertile, and humus-poor soils of tropical rain forests.

Changes in biodiversity aboveground can unravel this tight coupling, altering not only the subterranean environment but also the resources flowing into it. For instance, the nutrient quality, volume, or decay rates of litter vary according to the plants that generate it. Annuals die and yield their entire biomass to decomposition each year; perennials yield only a portion. Deciduous trees drop their leaves each fall; evergreens shed lesser amounts year round. Some plants also drop fruits, seeds, twigs, branches, bark, or wood. Some debris comes loaded with lignins,

waxes, tannins, and other unpalatable or even toxic compounds that require time and special talents to decompose. All of these factors influence the types of organisms present in the soil and the speed with
which they recycle organic matter and make nutrients available to support new plant growth. Likewise, conversion of large swaths of the landscape to intensive agriculture can completely decouple plant productivity from the process of decomposition and nutrient cycling. That's
because the harvesting of crops often eliminates organic inputs to the
soil, and fertilizer applications substitute for the natural process of nutrient release by soil organisms.

Any activity that alters what grows in the soil risks changing the way
the soil community functions: who does the work, how quickly, and
what products these organisms release. Unlike aboveground shifts, such
as the transformation of grasslands to desert, even radical changes in
the subterranean landscape tend to go unnoticed. Yet eventually,
changes underground manifest themselves above ground, as happened
in the clearcut areas on Washington's Olympic Peninsula, where the soil
community lost its capacity to support trees.

Scientists still know little about how the array of unseen and largely
unknown species in the soil contribute to its vitality. Only recently have
they begun to consider how loss of underground biodiversity might affect ecological processes critical to the maintenance of soil fertility.

The Underground Community

Fertile soil teems with life and activity; it is a subterranean micropolis
with billions of workers ceaselessly scavenging dead plant and animal
tissues from the surface world. It is hard not to think of these organisms
as a work force, a global service corps of rot and renewal. Yet each bacterial cell or mite or millipede, like every larger creature, works only to
nourish and perpetuate itself and, thereby, unwittingly plays a role in
generating larger ecosystem processes.

Across all biomes, from the tundra to the savanna, complex communities thrive in the top five to ten centimeters of the soil, where much of
the annual flow and exchange of carbon and nutrients takes place.[5] This
hubbub is not scattered uniformly throughout the soil but is concentrated in tiny hot spots of activity around the roots of plants, where microbes fuel themselves on sugars and organic acids exuded from the
growing root tips.

What happens deeper underground? For convenience, the U.S. Department of Agriculture Soil Survey declared in the mid-1970s that the lower boundary of the soil lies two meters below the surface, a depth that covers the rooting zone of major crops. But that designation is likely to change as our knowledge of the importance of soil processes grows. Soil scientists oriented to ecology rather than agriculture contend that the soil, meaning the strata where biology holds sway over the earth's crust, can extend many tens of meters down. Even at those depths, gases and organic acids released by living organisms contribute to the weathering of underground rock and the creation of new soil. Scientists studying deep-rooted plants and also water quality issues are increasingly interested in other ecological processes occurring many meters below the surface. But efforts to study the activities of life at such depths are just beginning.[6]

Even in the upper levels of the soil, probably fewer than 5 percent of the microorganisms have been isolated and named. In size they range from microscopic bacteria, algae, and fungi to macroscopic woodlice and earthworms. Many are lumped into broad functional groups according to the type of work they do rather than classified by distinctive individual traits.

Arguably most important, and certainly least censused of the soil organisms, are the microflora, the hordes of bacteria and fungi that are the most abundant and metabolically versatile organisms in the soil—indeed, on the earth. Identifying species and censusing populations of microflora in the soil has been all but impossible. Scientists know them largely by their sheer biomass. A single gram of soil from the root zone of plants, for instance, may contain a billion viable bacterial cells, representing four thousand to five thousand bacterial types. Some microbial ecologists suggest there are up to forty thousand types. The much-scrutinized soils of a tiny nature reserve in southwest England have already yielded twenty-five hundred species of fungi, and the count is increasing.[7] Microbiologist David Hawksworth, director of the International Mycological Institute west of London, estimates that there are between 2 and 3 million species of bacteria and 1.5 million fungi on the earth. Only 2 to 5 percent of these have been described and named.[8]

Each functional group among the microflora specializes in one of the multitude of chemical reactions needed to break down proteins, carbohydrates, and other complex organic molecules into simpler organic acids, then mineralize these into inorganic forms, such as the nitrate that plants can use. Some organisms excel at minute aspects of a larger task, such as the multi-stage processing of slow-rotting natural polymers like lignin, cellulose, and pectins. In any given patch of soil, there may be hundreds of species of fungi and bacteria working at a task in sequence like teams on a factory assembly line.[9]

Consider the processing of nitrogen, a vital plant nutrient. The major natural source of soil nitrogen is rotting organic matter. Certain groups of microbes process this material and release ammonium, which plants rapidly take up and incorporate into new proteins. Some plants, including many major row crops and grasses, however, absorb their nitrogen in the form of nitrate, which they are able to obtain because soils also harbor specialist teams of nitrifying bacteria that convert ammonium to nitrate. Before humans learned to synthesize nitrogen fertilizer, most new nitrogen supplies were drawn into the biological cycle through the work of another specialized group of microbes: nitrogen-fixing bacteria that convert nitrogen gas from the air into usable nitrate or ammonium. The nitrogen-fixers include some of the most economically valuable bacteria in the soil, such as the Rhizobia that live and work in nodules on the roots of peas, beans, and other legumes.

If all the microflora were to disappear, life on earth would quickly come to a halt, wilting and perishing into a never-to-rot compost pile spread across the land. Yet even these vital creatures do not work in isolation. The soil community includes ranks of invertebrate animals, some of which graze on bacteria and fungi, assuring the release and recycling of energy and materials tied up in even the smallest life forms. Many of these soil animals also speed the work of decomposition by shredding, consuming, digesting, and excreting organic debris, turning it into more accessible crumbs.

The tiniest of the soil animals are the microfauna, such as nematodes (roundworms) and protozoa that live mostly in water films around soil pores. Although some are plant parasites that cause serious crop damage, others graze on bacteria or fungi. They are joined by intermediate-sized soil animals, the mesofauna, which include highly specialized invertebrates, such as mites and springtails, that occupy air-filled soil pores and prey mostly on fungi. A square meter of soil may support populations of 10 million nematodes, 1 billion protozoa, and, depending on its organic matter content, 200,000 to 400,000 springtails and mites.

Largest and most noticeable of the soil invertebrates are the macrofauna: earthworms, ants, termites, millipedes, woodlice, beetles, insect larvae, and others. All are large enough to alter the physical structure of the soil and fragment the litter as they tunnel and feed, aerating the soil and forming channels for infiltration of water.[10] Each earthworm can ingest and excrete up to thirty-six times its own weight of soil each day. Earthworm populations may consume from ten to five hundred tons of soil per hectare per year, excreting it as dark and fertile castings.[11] In the tropics, ants are the chief earth movers, and they run a close second to earthworms in cold temperate forests as well.

Who Does What in the Soil

Exactly what most microbes in the soil do for a living—how they divide the resources, interact with one another, and contribute to ecosystem functioning—remains largely unexplored. So far, researchers interested in finding out what certain categories of soil creatures do have created controlled settings called microcosms. These are small-scale replicas of ecosystems that often start with a container of sterilized soil. By adding back only certain types of bacteria, fungi, or soil animals, and perhaps letting plants grow in the soil, scientists can see whether specific microbes change the rate at which nutrients are processed.

One question that's been asked is: What good are nematodes? Thanks to microcosm studies, scientists now know that these tiny grazers may be responsible for 30 percent or more of the nitrogen released to plants, useful work that has traditionally been attributed solely to the labors of bacteria and fungi.

Russell Ingham and others then in the lab of David Coleman at Colorado State University found that bacteria thrived in larger numbers when they were placed in a microcosm of grassland soils with their nematode predator. Blue grama grass grew faster, too, and initially took up more nitrogen when the nematodes were at work below. It turns out that bacterial cells contain more nitrogen than nematodes can use, so the feasting nematodes excrete a lot of it as ammonium wastes. Both the surviving bacteria and the plants clearly benefit from this extra nitrogen source.[12]

Similar results have been found in other soil types, from those of the pine forests of Sweden to those nourishing winter wheat fields in The Netherlands. For instance, wheat grown where both bacteria and bacteria-grazing protozoa were active grew significantly better than in soils with only bacteria present.[13]

Of course, in the real world, the action never involves just two interacting types of soil creatures, but rather a whole web of predators and prey.[14] These microcosm studies do show, however, that specific types of soil organisms, from nematodes to mites to millipedes, can alter the rate at which nutrients are cycled, at least across a minute patch of geography.

The more practical question for the human enterprise is: Do these life and death struggles, the appetites of specific nematodes or mites, have an irreplaceable impact at the scale of a crop field or landscape? Scientists don't have the final answer yet, but some believe that major processes, such as carbon and nitrogen cycling, are largely cushioned from these small-scale struggles by a great deal of redundancy. That is, the soil hosts many species with similar functions that can take up much of the slack when some of these soil workers are lost.

Some evidence for redundancy comes from agriculture. Consider, for instance, the harsh practice of soil fumigation. From the wheatlands of Australia to the strawberry fields of California, many farmers literally sterilize their soils with toxic fumigants before planting, a sledgehammer practice intended to kill fungal pathogens, root-feeding nematodes, and other diseases that build up in fields planted with the same crops season after season. Amazingly, fumigation often causes only transient effects on the major functions of the soil because many organisms recolonize or grow quickly from residual populations.

In the 1970s, for instance, soil scientists in the wheat-growing region of South Australia conducted a series of trials using the common agricultural fumigants methyl bromide and chloropicrin. The treatments eliminated virtually all soil fungi and 95 percent of the bacteria prior to planting the fields. Yet in less than a month, the researchers witnessed "a substantial recolonization" of the topsoil by fungi. Bacteria rebounded even faster, their numbers soaring within ten days to ten times as high as before the fumigation. The wheat seedlings planted immediately after the fumigation got an initial shot of fertilizer from the ammonium released by all the killed and rotting microbes. However, the fertilizer effect vanished within sixty days as the rebounding bacteria took control of the nitrogen cycle once again.

The result was a very different soil community than it would have been without fumigation. Bacterial types that had played a minor role before the fumigation gained dominance. Yet this highly modified assemblage was still able to restore functions in the disrupted system.[15,16]

Another team from Coleman's research group got a similar result using more selective chemicals—bacteriocides, fungicides, nematicides, and insecticides—alone or in combination to knock out specific types of soil organisms. The researchers found that when they knocked down nematode numbers bacterial populations increased. Soon, however, the number of protozoan grazers soared to take advantage of the new bacterial bounty. The complex tug and pull between microbial populations continued, so that even after seven months the community had not returned to its original makeup, but the nitrogen available to the grasses returned to previous levels. The microbial cast of characters shifted dramatically, and yet nutrient processing rates quickly returned to normal, as various microbes adjusted their populations to compensate for the lost workers.[17]

Whether the highly modified agricultural soils studied reflect the workings of natural soil communities is unknown, but there are reasons to be cautious about assuming that functions are buffered by redundancy in natural soils. Even though large-scale carbon and nitrogen cycles may be strongly buffered, loss of underground species diversity could be disruptive in ways not yet detected. That's because each set of workers may have traits and peculiarities that go unrecognized or may operate under certain optimal conditions, which affect how the work gets done.

A Canadian researcher found evidence that such subtle differences do exist. P. O. Salonius of the Canadian Forestry Service Maritimes Forest Research Center in New Brunswick took soil from beneath a stand of black spruce trees, dried it, and repeatedly diluted it in sterile water to reduce its density and diversity of microbes. He inoculated containers of sterile soil with these impoverished solutions and let their populations multiply until they matched the density of bacteria in the original soils. By monitoring respiration as they decomposed organic matter, he found that the more uniform populations weren't accomplishing as much as the same-sized complement of more diverse organisms. Apparently some task wasn't getting done as quickly or as well.[18] Do similar effects occur in crop fields, too? Do they contribute in subtle ways to declines in soil fertility and crop yield that neither farmers nor researchers can detect amid the more obvious problems of nutrient depletion and soil erosion?

Key Workers in the Soil

Although the overall impact of diminished soil biodiversity on the vitality of our soils has yet to be understood, several highly specialized—and therefore not readily replaced—organisms have been identified. The loss of these workers, or their introduction into new settings, can cause detectable alterations in soil processes.

Among these key soil workers are certain fungi that form intimate and mutually beneficial relationships with the roots of most plants. These are the mycorrhizae—literally "fungus roots"—best known for their gourmet truffles, mushrooms, and other edible fruiting bodies. Various types of mycorrhizal fungi grow around plant roots like a sheath, while others penetrate the cells of the roots. Either way they tap into the plant's own sugar or carbohydrate stocks for energy. In exchange, the fungal filaments serve as extensions of the roots, helping the plants absorb phosphorus and sometimes extra nitrogen, other nutrients, and water. Mycorrhizae also influence the soil community in the root zone, reducing the number of pathogens, sometimes increasing the population of nitrogen-fixing bacteria, and improving the structure of the soil.

Approximately 90 percent of plant species, from nara vines in the harsh Namib Desert to orchids, corn, conifers, and tropical forest trees form symbiotic relationships with mycorrhizal fungi, a partnership David Perry of Oregon State University and his colleagues refer to as one of the most "widespread and ecologically important symbioses in nature."[19] Some mycorrhizae are generalists, but others are highly specific in their choices of host plants, or even the life stages and conditions under which they partner up with certain plants.

Other critically important root symbionts are the nitrogen-fixers. A relatively small number of bacteria display this talent—that is, they carry the nitrogenase enzyme that allows them to turn gaseous nitrogen into ammonium. Various strains differ in how well they infect roots, how efficiently they fix nitrogen, and which host plants they prefer.

The power of a single species of symbiotic microbe, and its host plant, to directly alter the fate and functioning of an ecosystem is difficult to detect in healthy natural communities. However, the influence of a unique species trait can sometimes stand out starkly in a new setting.

Such is the case in parts of Hawaii Volcanoes National Park, where an exotic tree, *Myrica faya,* which carries nitrogen-fixing organisms in its root nodules, aggressively invades soils that form on recent lava flows. *Myrica,* a small evergreen native to the Azores and Canary Islands, was introduced to Hawaii, along with its microbial partners, in the late 1800s. Despite early efforts to keep *Myrica* from spreading, the trees have steadily advanced with the unwitting help of alien birds that carry their seeds. By 1985, *Myrica* had invaded 12,200 hectares in the park.

Recent lava flows and ash deposits are rich in some plant nutrients but low in nitrogen, a fact that prevents many native plants from colonizing the sites. An invasion by *Myrica,* however, profoundly alters soil development. Studies by Peter Vitousek of Stanford and his colleagues show that *Myrica* and its symbionts can quadruple the amount of nitrogen at a site, thus making more nitrogen available to other plants as well as itself. How this boost in fertility will alter the pattern of community development on these lava flows remains to be seen, but the ecological influence of *Myrica* and its nitrogen-fixing partners is likely to persist for a long time to come in the functioning of those soils.[20-22]

Other specialized talents in the soil community reinforce the notion that soil organisms from one site to the next cannot necessarily function interchangeably. Among these talents are the ability of some soil organisms to decompose organic materials that are tough and unpalatable.

It will come as no surprise to any gardener with a compost heap that pine needles and indeed, almost any type of woody debris, can sit in the pile for years without decaying. That's largely because of lignin, a complex polymer that combines with cellulose to stiffen the trunks and branches of trees. Lignin is so rot-resistant it often survives in fossils of woody plants. The higher the lignin content of the plant material, the longer it takes to decay. Toughest of all are the low-nitrogen, high-lignin needles and wood of conifers. A soil community with the right skills has no trouble at all with pine needles.

William Hunt of Colorado State University and his colleagues gathered bags of litter from a lodgepole pine forest in Wyoming, leaving some bags on the forest floor and some in a nearby meadow, then hauled the rest out to a prairie site in Colorado. They did a similar shuffle with bags of grasses from the prairie and meadow sites. More than a year later, the creatures in the meadow and prairie soils had made short work of the grasses, but the pine debris sat largely unchanged in its bags. On the forest floor, however, the pine litter rotted as quickly as the grasses. Clearly, the forest soils harbor specialists that are absent in prairie and meadow soils.[23]

In fact, healthy soils in conifer forests, and to a lesser extent most deciduous forests, are dominated by fungi, which can "reach" with their threadlike mycelial filaments into the thick debris layers that build up on the forest floor and eat away at the slow-rotting needles. In contrast, soils under grasslands and most row crops, where little debris collects, are dominated by bacterial cells, which are much more limited in their reach.

Using results at the level of litter bags and microbes to predict changes in soil functions on the landscape scale is not straightforward. Yet the message is inescapable: soils are specialized communities, many of them ill-prepared to sustain any type of crop or forest or other novel aboveground partnership forced upon them. Why, then, should dramatic shifts in the fertility and integrity of the soil that sometimes accompany changes in land use seem so surprising?

Clearcutting

One visible consequence of the Olympic Peninsula clearcut where young trees can't seem to get a toehold is the complete absence of fungal mats that once covered nearly a third of the forest floor. These mats—some of which look like ash from a campfire, others of which resemble bright white rubber bands, piles of yellow bird feathers, or even shreds of orange peels—are formed by the truffle-bearing mycorrhizae on which virtually all conifers rely. Their absence is but one sign of the wholesale change that has taken place since the clearing was logged.

Within the first year after the clearcut, in fact, Elaine Ingham of Oregon State University and her fellow researchers found that 90 percent of the fungi in the soil had vanished, including virtually all of the mycorrhizal mats. Other members of the soil community took a hard knock that first year, too. Bacterial biomass dropped two to three-fold; nematode numbers dropped two-fold; springtails and mites grew scarce; and a large pulse of nitrogen, apparently released by the decay of all those dead organisms, leached into the groundwater. In addition, three-fourths of the Douglas fir seedlings planted died in the first year. Five years later, more than 90 percent of the trees had given up. Clearly, the soil could no longer support a healthy stand of trees, evidenced by the fact that the soil community had shifted to a bacterial-dominated system more typical of grasslands than of conifer forests.

Across the road, on the site logged a year earlier, tree seedlings that were planted took hold and began to grow, indicating that the disturbed soil community could still nourish a forest.[24,25]

The Washington site is far from an isolated example. Further south in the high-altitude forests of southwestern Oregon and northern California, numerous clearcuts stand bare of trees two decades or more after logging despite four or five attempts to get new seedlings started.[26] One much-scrutinized site can be found at Cedar Camp on the slopes of the Klamath Mountains in southern Oregon. Today that site is covered with exotic cheatgrass, bracken fern, and an occasional manzanita bush. By all reckoning, it should be thick with twenty-five-year-old white firs. Four times since the slope was cut, crews have replanted fir seedlings; all have failed. In contrast, on nearby slopes charred by wildfires, young firs take root without a problem.

Obviously, some threshold had been crossed at the failed Olympic Peninsula site as well as Cedar Camp, beginning what David Perry of Oregon State and his team describe as a downward spiral of soil deterioration and associated seedling failure. Without tree roots, mycorrhizae, and other biotic elements to give it structure, the coarse-grained, granitic soil at Cedar Camp has taken on the texture of beach sand, and the proportion of bacteria to fungi is ten times greater than in the healthy fir forests nearby.[27]

Reconstructing which combination of logging or management practices triggered the failure at any given site has been difficult, and biologists like Perry and Ingham can't yet give firm answers. However, there are plenty of abuses to consider. For one, logging skids and other heavy equipment dragged too often across the landscape may compact the soil too much for air and water to penetrate. Also, fires set to clear the debris may sometimes burn too hot, devouring the protective carpet of needles and sterilizing the topsoil. Finally, the application of chemicals used to prepare the site for replanting—fumigants to kill root-rot fungi or herbicides to knock down competing vegetation—may prove too harsh for the soil community.

Aside from the direct impact of the herbicides, the act of killing off the shrubs and herbs that pioneer the clearing may disrupt the synergy between plants and soils. Perry points out that burning or clearcutting a forest obstructs the energy supply flow between the roots and the soil community. Organisms dependent on the sugar flowing to those roots— particularly mycorrhizae and nitrogen-fixers—are seldom equipped to survive without this subsidy. Unless the plant community rebounds after a clearcut, fire, or other disturbance and restores energy supplies to the soil, the balance will shift, favoring microbes that live on rotting organic matter. It's not the exact species composition of the soil community that's important (as mentioned earlier, the major nutrient cycles can often continue even with a shift in workers), but rather the functional partnership between the plants and soil organisms.

Once that partnership deteriorates, restoring the original plant community may not be possible. After several attempts to replant fir seedlings at Cedar Camp had already failed, Perry and Mike Amaranthus of the U.S. Forest Service tried inoculating the root zone of new seedlings with a cupful of soil from the root zone of healthy conifers, a step that doubled the growth and increased the survival of those conifer seedlings by 50 percent the first year. By the third year, the inoculated seedlings were the only ones still alive, although those scraggly survivors are not much healthier than the ones further north at the Olympic Peninsula site.

Ingham suspects that once the mix of organisms and the physical structure of the soil have changed, mycorrhizal fungi may not be adequate to support a forest. Even the hapless survivors on the Washington plot have mycorrhizae on their roots, although they may not be the optimal mycorrhizal species for getting young trees started. The larger problem is that once bacteria begin to dominate, they may turn the soil alkaline, rendering it more suitable for grass or crops than trees. In fact, Cedar Camp is now an annual grassland.

One clear caution for the future has come out of these studies. As climate change shifts the territorial ranges of many plants, Perry and his colleagues see a real danger that current vegetation could die off before new species can colonize a region, thus disrupting the plant–soil linkages. Or, some impoverished sites could be captured by weeds—those "mobile opportunists that invest relatively little in soils." The more diverse a plant community, the more likely that some nurturing linkages remain intact during the transition.[28]

Plantations of Pine

Ironically, while foresters sometimes struggle to get conifer seedlings to come back on soils that once supported rich old-growth conifer forests, timber companies find it easy and profitable to convert many of the world's hardwood forests to conifer plantations. In many temperate forests, deciduous trees such as beeches, oaks, maples, and birches are being replaced by tree farms of spruce, pine, and fir. The incentive is financial: Pines grow faster initially and so produce more biomass for the same nutrient input than hardwoods do over the short term. Pines are thus ideal for those who want to chop and regrow pulpwood on short time cycles.

Across Europe, the conversion of deciduous forests to evergreens has been underway for a century. The forests of Germany, once 90 percent broad-leaved trees, are now more than 80 percent Norway spruce, Scotch pine, and other non-native conifers. In eastern North America, large tracts of mixed hardwood forest have been cut and replanted to monotonous stands of loblolly, white, and jack pines. In Japan, diverse forests of oak and maple have largely been replaced by dense, uniform swaths of Japanese cedar (with public health consequences, as described in Chapter Three). In the southern hemisphere, many rich natural systems, from the once impenetrable beech forests of southern Chile and Tierra del Fuego to the eucalyptus woodlands of Australia, have been razed to make way for commercial stands of Monterey pines introduced from California.[29]

These wholesale shifts in the aboveground community require a great deal of human assistance at first. Foresters must usually fight back the regrowth of hardwoods and understory vegetation for several years to keep them from outcompeting the conifer seedlings. That's because the soil harbors a rich legacy from decades or centuries of supporting deciduous forests. The rain of nutrient-packed and highly palatable leaf litter from these trees encourages a very active soil community that rapidly converts the debris to thick layers of humus. The result is a highly fertile soil, stocked with quickly available pools of nitrogen, as well as alkaline minerals such as calcium and magnesium.

Conifers are specialists on infertile soils, ill-equipped to compete with oaks or maples on overly rich sites. However the conifer canopy, once it closes, works as well as a herbicide in killing off any deciduous seedlings or understory shrubs. Year-round, the evergreen canopy blocks sunlight from the forest floor and intercepts a larger proportion of the rainfall than deciduous trees, keeping the soil below cooler and drier. Spongy carpets of needles pile up faster than the soil community can rot them, forming a thick, acidic debris layer across the surface.[30] Little but moss thrives in the understory.

Conifer litter not only contains materials that help acidify the forest floor, but can also harbor toxic compounds, such as tannins, that make it difficult for competing plants to capture any resources. Indeed, when Robert Northup of the University of California, Davis, and his colleagues examined stunted pines growing on a series of coastal terraces in Northern California, they found that pines on the most acidic and infertile soils had the highest levels of tannin in their litter. These tannins help block the efforts of soil organisms to break down the litter and release its nitrogen in a form most plants can use. This may be the pines' own way of hogging scarce resources. It turns out that mycorrhizal fungi on the roots of the pines make enzymes that can break down this material and recapture the nitrogen for the trees. Thus the pines can effectively monopolize and horde within their own tissues most of the key nutrients in those harsh soils.[31,32]

The conversion to conifer plantations, therefore, represents a loss not just of tree diversity but of the flowering plants and shrubs that abound in deciduous forests, as well as the animals that rely on a lush understory for food and habitat.

The conversion also results in a gradual loss of nutrient capital from the soil. That's because the microbes, accustomed to a fast-paced life with bountiful nutrient inputs from the hardwoods and a generous uptake of liberated nutrients by the trees, continue to recycle nitrogen at their usual pace, even after the switch to conifers. Young pine trees may get an initial growth spurt from this largess, but they cannot take up all of it. So a massive leaching of nitrates to groundwater and stream flow often ensues as microbes proceed to break down the humus stockpiled in the soil.

The nitrates do not go alone, however. When these negatively charged materials dissolve and leach away, they tend to pull along positively charged minerals, such as calcium and magnesium. The loss of these alkaline minerals impoverishes soil nutrient stocks and helps to acidify the soils. In some places where acid soils have become troublesome, especially in Europe where the effects of air pollution and acid rain enhance the soil changes caused by forest conversions, land managers have tried liming the soils to reverse the pH. Unfortunately, liming can have other unwanted effects, including spurring more leaching of nitrates into water supplies.[33]

Tropical Soils

It seems counterintuitive that a land thick with ancient forests and lush vegetation could sustain such plant wealth on soils too poor to support crops. Certainly when nineteenth-century explorers such as the British naturalist Alfred Russel Wallace beheld the tropical rain forests of Amazonia, they equated greenness with fertility. Wallace wrote, "I fearlessly assert that here, the primeval forest can be converted into rich pasture-land, into cultivated fields, gardens and orchards, containing every variety of produce, with half the labor and, what is more important, in less than half the time that would be required at home."[34] Certainly, the perceived fertility of tropical soils was an understandable illusion then, but as time has shown, it was a grievously misleading one.

The truth is that soils in many parts of the tropics are old, highly weathered, acidic, and mineral-poor. When forests are cleared and these lands sown to row crops or pasture, the soils deteriorate quickly. Recognizing that fact, governments and would-be farmers should be doing everything they can to protect tropical soils. Yet tropical forests around the world are being razed at the rate of 15.4 million hectares per year—three times the land area of Costa Rica. Worse, only about half of that lost forest actually expands the amount of land in crop or cattle production. The rest simply replaces the vast stretches of worn-out land that are abandoned each year, often after only one to three years of agricultural use. Thus, Richard Houghton of the Woods Hole Research Center in Massachusetts points out that the rate of tropical deforestation could be halved without even slowing the current growth of cropland and pastures just by instituting sustainable agriculture, protecting the soils, and ending land abandonment.[35,36]

What explains the deceptive lushness of tropical rain forests? The answer, it turns out, relates more to the species-rich aboveground community than to any attributes of the soil itself. These species-rich forests literally nourish their own diversity, creating a far richer aboveground community than the tropical soil alone could support. Nutrient levels in many tropical soils are so low that lush plant communities couldn't survive if they allowed minerals to be leached by the three or more meters of rain that drench the region each year. So the enormous diversity of species creates an almost closed system when it comes to cycling nutrients. Trees in most tropical forests, for instance, put out a dense mat of fine roots on or just below the surface to retrieve nutrients as soon as they are released from the litter and to capture rainwater quickly, before it can filter down to groundwater.

In addition, the forest canopy protects the soil from chemical weathering, which proceeds three to six times faster in the tropics than in temperate regions. The canopy also captures a large proportion of rainfall as stemflow. As the water runs across leaves and stems and down the trunk it picks up nutrients from the excrement of a host of animals living in the canopy, along with substances leached directly from the tree's own tissues. Chemical analysis shows stemflow water in central Amazonia is richer in dissolved mineral nutrients than the original rainwater: nitrogen compounds average fifteen times higher, phosphorus thirty times, potassium sixty-five times, calcium twenty-five times higher.[37]

Clear and burn these trees, however, eliminate the forest community, and the soils deteriorate quickly. The ash of the burned forest is usually alkaline enough to reduce the acidity of the soil for a time; the ash also contains enough nitrate to provide a good boost for the first maize or grass crop, or weeds that recolonize the site. Within weeks, though, nitrate supplies drop as ash erodes and leaches. The meager nutrient supplies stockpiled in the soil are also depleted quickly. Just as in temperate forests, microbes feeding on soil organic matter yield up its nitrogen, phosphorus, sulfur, and other stockpiled resources to the rapidly growing crops or weeds. Yet the reservoir of organic matter in tropical soils is scant and short-lived compared to that of temperate deciduous forests. Without the trees, abundant rain and sun also injure the soil. The crops or weeds cannot take up all the rain that falls, so erosion and leaching quickly deplete the remaining nutrients in the surface layers. Without the constant shade of the forest canopy, the soil gets a year-round blistering from the tropical sun.

In traditional slash-and-burn agriculture in the tropics, farmers clear and crop small plots for only a few years, then move on, leaving these areas fallow, commonly for at least six years, although sometimes for decades. If the soil of these plots has been too degraded—crossed some little-understood threshold like those forest clearcuts did in the U.S. Pacific Northwest—it may take centuries before natural processes can restore soil vitality. Yet, if abandoned fields have not been too heavily degraded, they are usually recolonized quickly during the fallow period by a lush and vigorous succession of plants that restore soil fertility and productivity. This natural restoration process intrigued Jack Ewel, now with the U.S. Forest Service Institute of Pacific Islands Forestry in Hawaii, because the regrowth takes place on soils incapable of supporting another monoculture crop without heavy infusions of fertilizer.[38]

To help understand the role of plant diversity in this restoration process, Ewel's team set up a series of experimental plots in Costa Rica. As described in Chapter Two, the team cut and burned the trees, then created three types of high-diversity plots. On one type, they allowed the natural successional vegetation to spring up, a strategy that after five years resulted in a community of more than one hundred species. On another type, the researchers assembled an equally diverse array of plants, matching the natural succession "vine for vine, herb for herb, shrub for shrub, trees for tree, and epiphyte for epiphyte," but using a completely different set of species. In yet a third type of plot, they left the natural vegetation to regrow but also scattered thousands of extra seeds. In the end, these plots held some two dozen more species than the natural successional plots that had not been enriched. Ewel's team also planted a sequence of monocultures timed to mimic the architecture of the plants recolonizing the high-diversity plots: maize the first two seasons, then cassava shrubs, and last a fast-growing tree.

Finally, one plot was left bare of all vegetation and kept that way for five years by hand weeding. Arduous at first, the task became quite easy as the study progressed; after years of exposure to the elements and without plant roots and litter to nourish microbes and replenish nutrients, the soil became so degraded that few plants sprouted despite a continuous supply of seeds raining down from surrounding vegetation.

As expected, the monoculture plots suffered a greater loss of soil nutrients during the five-year trial than the high-diversity plots. But nutrient depletion declined dramatically as the monocultures progressed from maize to shrubs to trees, at which point the plots showed only modest nutrient losses. After five years, the team also found little difference in soil fertility among the three highest-diversity plots. One hundred species seemed to sustain the soil as well as 125, and the artificially assembled community of 100 did as well as the natural assemblage. Somewhere between 1 and 100 species, the benefits of diversity—at least for nutrient cycling—peaked. The fact that 1 tree species was almost as efficient at maintaining soil fertility as 100 trees, shrubs, and herbs has important implications for the design of sustainable agricultural systems.

Trees and perennial shrubs "are the key to the site-restoring powers of fallow vegetation in the humid tropics because of their deep, permanent root systems," Ewel concludes. Annual crops are too short-lived to develop extensive root systems. Shrubs and trees, on the other hand, develop roots that thoroughly exploit soil resources, stretching into the subsoil to pump deeply leached nutrients back up to the surface. These roots also trap and recycle the nutrients that fall with the rain. In fact, the ability of these long-lived root systems to maintain soil fertility is a key reason why the most sustainable crops in the humid tropics are perennials: bananas, cocoa, rubber trees, and oil palms.

Even monocultures of such perennial crops maintain soil fertility fairly well over the short term. Yet, as the study in Costa Rica showed, the difference between one species and no plant cover at all is dramatic. Ewel and his colleagues point out there's a substantial risk that a single species may fail at some point, leaving the soil exposed. A multi-crop system not only maintains soil fertility most effectively but also provides insurance, increasing the likelihood that even if one or more species fail, something will be left to protect the soil and sustain those linkages between the aboveground and belowground communities.[39]

Dung Beetles

Another organism that contributes significantly to soil fertility is the dung beetle. Also known as scarabs, these insects can be found worldwide, aggressively recycling the waste of their fellow animals. They range from a few millimeters to nearly fist-sized. Some specialize on the droppings of a single species; others are generalists. When fresh droppings hit the ground, scarabs converge by the hundreds or even thousands, carving up the mess, eating some and burying the rest, laying their own eggs inside buried dung balls. The early Egyptians were so impressed with the beetles' labors that they depicted their sun god as a scarab, pushing the fiery orb across the heavens each day like a beetle rolling a dung ball across the earth. Although scarabs have slipped from the pantheon, they still render irreplaceable services: dispersing dung, aerating the soil with their tunneling, and assuring that most of the nitrogen in dung pats is released into the soil for fertilizing plant growth rather than being lost to the atmosphere.

The importance of dung beetles was demonstrated when the first English colonists arrived in Australia in 1788 with seven cattle in tow. At the time, Australia was home to 250 species of scarab beetles, none of which proved up to the task of dispensing cow dung. The beetles were well adapted to processing the relatively dry, fibrous, and modest-sized droppings produced by kangaroos, emus, and other indigenous animals. However, the dung pads produced by cattle were enormous, wet, and unmanageable, and eventually there were vast quantities to be man-

Dung beetles or scarabs contribute to soil fertility by dispersing and burying dung and aerating the soil with their tunneling.

aged. Within two centuries, Australia had 30 million cattle. Producing a dozen dung pads apiece a day, these cattle could literally pave over 2.4 million hectares of pasture with unprocessed manure in a single year. This "fecal pavement" can take months, or in some regions, years, to rot or weather away, in the meantime locking up nutrients and hindering the growth of pasture grasses. The mounds of dung also provide fertile breeding grounds for the fierce Australian bushflies that make outdoor life miserable for people and cattle.

To control this problem, Australia's Commonwealth Scientific and Industrial Research Organization first turned to Africa, which boasts more than two thousand species of dung beetles, some of them adapted to process the formidable output of elephants and other large mammals.[40] Eventually more than two dozen scarabs from Africa, Asia, and Europe were brought to Australia and released. Today, some thirty years later, seventeen of the foreign species have successfully established residence in the pastures of New South Wales. Researchers monitoring dung beetle and bushfly interactions on fresh cattle dung have found that high numbers of imported beetles seem to be suppressing the bushflies.

More important, they are keeping pasture soils nourished with nitrogen and accessible to grazing.[41,42]

Earthworms

The more common problem caused by intensive livestock and agricultural operations is not rendering the work of local soil creatures inadequate but creating such harsh working conditions that their populations are greatly reduced or eliminated. Clearing, burning, tilling, compacting the soil with heavy machinery, applying pesticides and herbicides, and reducing plant diversity in crop fields all take a drastic toll on the underground community. One of the soil animals frequently hard hit is the earthworm. Yet one expert asserts that the role of earthworms "in promoting soil fertility makes them probably the most important, in terms of their relevance to productivity, of all animal groups that share with mankind the earth's land surface."[43]

At one site in the Peruvian Amazon, a team led by Patrick Lavelle of the Laboratoire d'Ecologie de l'Ecole Normale Superieure in Paris counted more than forty-three hundred individual soil animals, including earthworms, termites, millipedes, spiders, mites, ants, and beetles, for every square meter of the floor below an undisturbed rain forest. Clearing and burning that forest and replanting it with maize destroyed more than 80 percent of those creatures, especially the earthworms. A clear demonstration of the beneficial impacts of earthworms came when Lavelle and his colleagues later reintroduced them along with various mulches to some enclosed plots in the maize field. Grain yields increased anywhere from 23 to 60 percent in the first season.[44,45]

In New Zealand, earthworms were introduced to one isolated pasture and within five years their population had grown to 1,150 worms per square meter. Their feeding, burrowing, and casting activities reduced surface litter, increased the organic matter in the soil, doubled water infiltration, and, within a decade, boosted grass production by 28 percent. Similar beneficial results have been reported for worms at work in Dutch orchards, French pastures, untilled German cornfields, and fallow croplands on the Nigerian savanna.[46]

Rising CO_2 and Nitrogen Overloads

Even without the use of plows, chainsaws, or fertilizers on lands untouched by direct human disturbances, human activity is having a pro-

found impact on natural nutrient cycles and the interdependence between plants and the soil community. Every time people drive cars, fire
up power plants or factories, or burn forests, we are releasing gases and
particulates that impact the functioning of ecosystems worldwide.

The burning of carbon-rich fossil fuels, such as coal, gasoline, and oil,
and the clearing and burning of forests in the tropics have helped drive
carbon dioxide (CO_2) levels in the atmosphere up from about 270 parts
per million two centuries ago to 350 parts per million today. At current
rates, CO_2 concentrations could double to 700 parts per million by 2050.
Even freezing carbon emissions at current levels would only slow, not
halt, that rise.

Most of the worry over rising CO_2 stems from the fact that it is one of
the so-called greenhouse gases. These gases trap outgoing heat from the
earth, raising global temperatures. Debate still rages over how much average temperatures will rise as greenhouse gases build up, and indeed,
whether greenhouse-induced global warming has already begun. Without question, atmospheric CO_2 levels are rising, and that trend, regard-
less of its larger implications for climate, has the potential to seriously
disrupt plant communities.

Experiments in pots, greenhouses, and growth chambers show that
high CO_2 often causes plants to grow faster and larger, at least for a time.
For reasons not completely understood, the effect doesn't last. For one
thing, plants still need nitrogen, phosphorus, and other nutrients; if any
of these are in short supply, the potential fertilizing effect of high CO_2
has little impact on plant growth.[47]

Another consequence of extra CO_2 is its potential impact on nutrient
cycling by the soil community. Plants growing in high CO_2 usually pack
proportionately less nitrogen-rich protein and more carbon-rich starch
into their leaves and thus have less nitrogen to offer the soil microbes.
It remains to be seen whether the net impact on the soil community will
be negative or positive. That's because plants raised in high CO_2 often
allocate much of whatever increased growth they experience to roots
rather than to stems and leaves. With more roots to exploit the soil, and
more sugars to nourish their mycorrhizal fungi, nitrogen-fixers, and
other soil microbes, some researchers think plants just might spur
greater soil fertility.[48]

Carbon is only half the story. Indeed, thanks to the invention of automobiles and industrial fertilizer production, humans now fix and make
available to the living world more atmospheric nitrogen than all natural
processes combined.

Peter Vitousek of Stanford points out that land-based ecosystems naturally fix about 100 billion kilograms of nitrogen a year; marine ecosystems add 5 to 20 billion kilograms; and lightning fixes another 10 billion. That's a maximum of 130 billion kilograms of nitrogen drawn from the air into circulation through soils, plants, and animals by natural processes. In contrast, human beings fix some 135 billion kilograms a year. Of that amount, more than 80 billion kilograms comes from fertilizer production. Another 25 billion kilograms is converted into nitrous oxide during the operation of internal combustion engines. Some 30 billion kilograms can be attributed to legume crops, such as peas and beans, that produce more nitrogen than natural vegetation on the same lands would have generated. Moreover, draining wetlands, burning forests, and other human activities release enormous but still unquantified amounts of nitrogen that would otherwise be locked into long-term storage.[49]

Much of this human-liberated nitrogen including nitrous oxide released in automobile exhausts or emitted from over-fertilized farm soils, and atmospheric ammonia, which emanates from animal wastes and farm soils, pollutes the air. Some of this nitrogen is deposited as windborne dust, showering even the remotest wild lands with extra fertilizer. Some, particularly nitrous oxide, is converted along with even more problematic sulfur compounds into acids that rain onto the land.

Direct damage to living organisms from acid rain, of course, has long been a serious concern. For several decades, acidification of lakes and soils has been apparent in central Europe and parts of the northeastern United States. Across western Europe, the proportion of forest soils with a pH more acid than 4.2 (7 is neutral) increased from less than 1 percent in 1960 to 15 percent by 1988, a process driven both by acid rain and the conversion of natural deciduous forests to conifers described earlier.

Coincident with that change in soil chemistry, researchers have reported declines in the abundance and species diversity of mycorrhizal fungi across Europe, from the oak and pine woodlands of The Netherlands to the spruce and beech forests of Germany and alpine regions of Austria and Czechoslovakia. Since the early 1970s, European collectors of boletes, chanterelles, truffles, and other mycorrhizal products have been troubled by a growing scarcity of these delicacies in the forests of the continent.[50]

Although acidification may be injuring mycorrhizae and other soil organisms and plants directly, scientists have recently begun to realize that excess nitrogen also may be damaging natural ecosystems indirectly—that is, by over-fertilizing them. For one thing, it's well known that fertilizing a field sharply reduces the diversity of plant species. Released from a hardscrabble existence by extra nitrogen, a few fast-growing plants will expand rapidly and outcompete their less aggressive neighbors for space and resources. In heathlands and grasslands of The Netherlands and other formerly nutrient-poor ecosystems across western Europe, for instance, scientists blame an increase in nitrogen deposition for the "great losses of biological diversity" that have occurred over the last few decades.[51]

Recently, scientists have become aware that there may be another, more subtle, impact of this rain of excess fertilizer than shifts in plant competition: They believe that the fertilizer effect of nitrogen may reach a saturation point, when more nitrogen is available than the plants and soil microbes can possibly handle. The resulting disruption of plant–soil linkages may lead to ecosystem degradation even before the damaging effects of acidification reach critical levels.

How could nitrogen overload threaten the health of a forest community? First, excess nitrogen in the form of nitrates would seep into streams, decreasing water quality. At the same time, the nitrates would carry away with them alkaline minerals, such as calcium and magnesium, leaving the soil less buffered against acidification and also deprived of vital nutrients. Acidic conditions also mobilize aluminum ions in the soil, allowing these to build up to potentially toxic concentrations that can harm tree roots and perhaps inhibit the growth of mycorrhizae. That's one mechanism besides direct acid rain damage by which excess nitrogen could be damaging mycorrhizae in European forests. Another possible explanation is their weakened relationship with the trees. With externally supplied nitrogen so abundant, there's evidence trees expend less energy to support symbionts like mycorrhizae and nitrogen-fixers.

Breaking these ties could be a costly mistake for the trees, however. Mycorrhizae, as noted earlier, often provide protection from root diseases and other services besides nutrient gathering. Their decline, combined with the stress of soil acidification, could eventually hinder tree growth and productivity. As the trees declined, they would likely prove more vulnerable to frost, disease, or insects. Unhealthy trees, in turn, would take up even less nitrogen from the soil, reinforcing a self-perpetuating decline.[52,53]

Evidence of just such a downward spiral was reported recently by Walter Durka and Ernst-Detlef Schulze of Bayreuth University in Germany and their colleagues. By developing a way to track nitrates that originate as air pollution, these researchers were able to show that a great deal of the nitrate in streams draining from heavily polluted spruce plantations in Bavaria had seeped into groundwater without ever being taken up by plant roots or microbes. In heavily damaged and dying stands of spruce, virtually all of the atmospheric nitrate passed directly into mountain springs without ever being drawn into the biological cycle. Even in apparently healthy stands or those showing only slight decline, 16 to 30 percent of the nitrate passed untouched through the soil. Clearly, nitrate levels exceed the capacity of plants and microbes to absorb them, even in an apparently healthy forest. Researchers believe this loss of control over the nitrogen cycle may signal the onset of pollution-driven forest decline long before its effects show up as yellowed pine needles and thinning tree canopies.[54,55]

Once again, humans have unwittingly altered the synchrony between plants and the soil, disrupting the work of the underground community in unseen ways that eventually render it unable to support the vital aboveground processes: the productivity of forests, grasslands, and crops; the provision of pure water; the capture and recycling of valuable nutrients; and the maintenance of diverse and stable natural communities.

CHAPTER SIX
Of Plants and Productivity

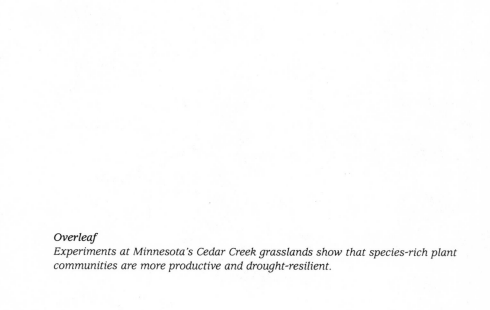

Overleaf
Experiments at Minnesota's Cedar Creek grasslands show that species-rich plant communities are more productive and drought-resilient.

*M*aps of the earth's major vegetation zones show a green swath of deciduous forest stretching across the northern temperate regions of Asia, Europe, and eastern North America. That green band has had a tenuous existence, however. Over the past thousand years, great tracts of these forests have been cleared for towns, roads, farms, pastures, and industries, and that's only the most recent upheaval. Amid the remnants of virgin forest that still dapple the landscape, evidence exists that agents far more powerful than plows and bulldozers have altered the forests time and again, long before modern humans stamped their presence on the land. These agents left behind an invaluable natural laboratory for ecological study, providing critical evidence to scientists about the relationship between species richness and plant productivity.

If a modern human had been able to visit Wisconsin or Poland or northern China at the dawn of the Pleistocene a million years ago, she probably would have noticed much the same diversity of trees and understory shrubs throughout the temperate forests. Then the great ice ages began. Across Europe and North America, the repeated advances and retreats of glaciers took a heavy toll. The forests of Europe and western Asia suffered most as the ice sheets advanced southward, trapping the retreating flora against the east–west bulwark formed by the Pyrenees, the Alps, and the Caucasus mountains. The forest species of eastern North America had a better chance of survival. No impassable barriers blocked their retreat as the glaciers crept down the north–south line of the Appalachians. When the glaciers began their last retreat twenty thousand years ago, surviving tree species recolonized to create the temperate forests as they appear today, each moving north at its own pace as the ice sheets melted.

The reassembled deciduous forests of North America ended up with 157 species of trees and shrubs, those of Europe only 106. In East Asia, which largely escaped the scouring ice sheets, deciduous forests still boast 876 species. One puzzle for scientists looking at this widely divergent number of species in various forests is this: Do five or eight times as many kinds of oaks, beeches, maples, ashes, lindens, willows, poplars, alders, birches, chestnuts, elms, and understory shrubs make a forest more productive than one with just as many trees but drawn from fewer species?

Such a question deserves careful appraisal. After all, plant productivity undergirds the earth's food webs. Measured as the total annual output of plant material from an ecosystem, primary productivity is visible in the lushness of leaves, stems, branches, roots, shoots, seeds, fruits, and woody growth. In short, productivity represents the growth that supports all animal consumers, their predators, and ultimately the myriad creatures of decay. Determining the relation between species richness and productivity is vital both in the face of biodiversity losses in natural ecosystems and in the context of modern agriculture. In virtually all human-managed systems, from crop fields to plantation forests, people have reduced the number of plant species far more drastically than any glacier ever did.

Researchers tackled the issue by examining productivity data from forty-two intensively studied temperate forest sites around the world, ranging from highly diverse stands to those dominated by only a few tree species. Their findings are reassuring, at least in the short run: Productivity does not seem to decline as the number of plant species in the forest drops.[1] The few studies and evaluations available from other types of ecosystems—tropical forests, temperate grasslands, tropical savannas, and even croplands—also support the view that beyond some apparently low threshold number of plant species, adding more does little to enhance the annual bounty.

Thus, in the near future, productivity may be one of the least disrupted ecological services as species vanish, and even an increasingly impoverished earth is likely to remain green enough to provide food, fiber, and fuel for some time to come. That's not surprising. After all, capturing carbon through photosynthesis is something every plant from algae to redwoods can do. Yet productivity may be a deceptively poor measure of long-term ecological health.

Although it's true that every plant can photosynthesize, none is equipped to produce prolifically on every soil type or slope, through every season, in the face of drought, flood, insect attack, windstorm, or fire. At what point will the productivity and other ecological services of the North American temperate forests falter, as introduced scourges like chestnut blight, Dutch elm disease, and white pine blister rust kill off major tree species one by one? How many backup species will it take to maintain stable productivity in the face of major disruptions, such as air pollution or global warming? How would the European forests have fared if they had faced the glacial epoch with today's 106 species—or single-species conifer plantations—rather than 800-plus?

Farmers, who can often maximize production in any given season by growing a single crop, have a skewed view of productivity. That's because much of their success reflects commodity market supports, subsidized water supplies, chemical fertilizers and pesticides, and other economic artifacts of industrialized agriculture. The real payoff of diversity, it seems, comes with time. In both agricultural and natural systems, evidence indicates that diversity at the genetic and species levels helps to stabilize productivity over the long term, allowing the community to capture and use available resources efficiently and resist and rebound from the vagaries of climate and other shocks. Our definition of healthy natural ecosystems—and increasingly, of agricultural ecosystems as well—should center not on maximal productivity in a single year but on sustainable productivity over time.

One conservative 1986 analysis calculated that our species already uses or co-opts 40 percent of the potential net primary productivity of the entire biosphere. The amount we use directly as food, fiber, timber, or fuel represents about 7.2 billion metric tons of organic matter a year, or 3 percent of the earth's output. Add the net primary productivity of croplands, gardens, golf courses, lawns, parks, and other areas devoted entirely to human activities, plus the biomass fed to livestock or burned to clear land, and the total rises to 19 percent. Finally, throw in the productive capacity lost when open land is paved for cities and highways, forests are converted to less productive cereal crops or pastures, and land is degraded through processes such as desertification. The result totals 58 billion out of nearly 150 billion metric tons of organic matter. Even that calculation ignores human impacts such as pollution, acid rain, and atmospheric change, which greatly affect the remaining 60 percent of the earth's productivity.[2]

Obviously, sustaining the earth's fruitfulness is critical to human survival. In recent millennia, expansion of croplands and pastures to feed a burgeoning human population has been a major driving force changing the landscape and eliminating habitat and resources for other species. Indeed, nearly one-third of the earth's land surface today is de-

voted to agricultural uses. More disturbing is that half the currently cultivated lands have been added in the past ninety years, a pace that reflects the increasing rate of human population growth. In the tropics,
agricultural lands have doubled over the past fifty years. Any hope of significant expansion to meet future needs bears little basis in reality. Of
the two-thirds of the earth's land surface that remains uncultivated, half
is rock, ice, tundra, desert, and human settlements unavailable for agriculture. The other half hosts all of the world's surviving forests, woods,
shrublands, and grasslands. Most of this land is marginal for farming because of its poor soils, steep slopes, or low rainfall, so even if all of it
were sacrificed to agriculture, humans could not expect to double food
production.[3] In fact, most of this remaining land is more productive and
provides a wider range of ecological services in its natural or semi-man-
aged state than it would yield under the plow.[4]

Not all plant communities are equally efficient at converting light into
usable carbon compounds. A tropical forest may convert 1 percent of
the solar energy it receives to plant tissue; a tall-grass prairie only one-
tenth and a desert one-twentieth of 1 percent. An acre of corn is extremely effective during the growing season, using 1.6 percent of the
solar energy it receives; but averaged over a year, the efficiency of a
cornfield at using light drops to less than one-half of 1 percent.[5]

Under optimal circumstances, high intensity agriculture can rival the
net primary productivity of the earth's most bountiful natural systems—
tropical rain forests, coral reefs, swamps, and estuaries—growing thirty-
five hundred grams or more of plant matter per square meter each year.
But in actual practice, farmers around the world fall wide of that mark,
so much so that the global mean for cropland productivity is only 650
grams. That's about one-half of the mean for temperate forests, and less
than one-third of the mean productivity of tropical rain forests, coral
reefs, and swamps. Tragically, it is these high-productivity ecosystems
that are most threatened, directly or indirectly, by humans making way
for agriculture.[6]

Low-Diversity, High-Input Agriculture

Much of modern agriculture, particularly in the northern temperate
zone, reflects the eggs-in-one-basket strategy of wringing high yields
from extremely low-diversity systems, often comprised of a single genetic strain of an annual cereal crop, such as wheat. (Yield refers to the

fraction of productivity that humans value and use. In the developed world yield may refer only to the grain of wheat, while in traditional societies it may include the stalks, which are used for livestock forage or fuel.)

The result of such a strategy is that literally a handful of species feed the world. More than 90 percent of the world's food is supplied by fifteen plant species, and nearly two-thirds of that total comes from just three grains: rice, corn, and wheat. The rationale for such simplification has more to do with economic considerations, such as subsidized markets for commodity crops or cost-effective scheduling of mechanized farming and shipping than with biological principles. Indeed, this approach calls for obliterating the natural plant community from a patch of land, wiping the slate clean to make way for a reconstituted system designed to perform only one function: maximal output of a single type of profitable biomass.[7]

It can be expensive and labor-intensive to maintain a single-purpose system that cannot carry out the other essential functions of a self-sustaining ecosystem, such as restoring soil fertility, preventing erosion and nutrient leaching, capturing rainfall, and supporting the natural enemies that control insect pests. In addition, monocultures are critically sensitive to annual variations in rainfall, nutrient supplies, competition from weeds, and pest and disease outbreaks. Intensive management practices such as tillage, irrigation, and applications of fertilizers, herbicides, and pesticides, are designed to suppress or smooth out variables like these, though over time they, too, can cause drastic losses.[8]

Despite intensive management—or often, because of it—the high yields achieved by crop monocultures can level off and decline. Crop yields suffer when pests grow resistant to chemical pesticides and when soils deteriorate without a steady input of carbon from plant litter or they fall prey to erosion and salinization. Salinization is a serious problem, as mentioned in Chapter Four. One-third of the world's food is grown on artificially irrigated lands, and one-third to one-half of those lands already suffer moderate to serious salt problems. In the Australian wheat belt, where the elimination of shrubs and trees has led to extensive salt intrusions and waterlogging, wheat yields have leveled off over the past twenty-five years.

Indeed, around the world, the soaring yields of grain achieved by the agricultural intensification that followed World War II have turned sluggish during the past decade. Wheat output made no gains during the

1990s. Rice is up slightly, but only after a dramatic drop in the overall growth rate of rice yields across Asia during 1980s. The Worldwatch Institute believes part of this sputtering growth in yields can be blamed on loss of soil fertility due to increasing soil degradation.[9,10]

Another aspect of modern agriculture that worries ecologists is the narrowing of genetic diversity within each crop species. In the United States in 1982, for instance, a single variety of potato, the russet Burbank, accounted for 40 percent of all potatoes—fully 95 percent in the potato-growing state of Idaho. The "green revolution" of the 1960s provided high-yielding but genetically uniform strains of rice, wheat, and other crops to farmers worldwide, encouraging them to abandon hundreds of traditional cultivars. One of these rice varieties, IR36, was planted across 70 percent of the rice paddies in the Philippines in 1981–1982 and is now the dominant rice variety in Indonesia, Vietnam, and India.

Capable of dramatically improving yield in the short run, IR36 produces two to six times the harvest of traditional rice varieties. But, while the traditional cultivars or landraces are typically low in yield, they are robust in their resistance to pests and disease. This robustness is notably lacking in the new hybrid crop varieties bred solely for high yields. In fact, the very name is telling: IR36 represents the thirty-sixth strain released by the International Rice (IR) Research Institute, each one bred to resist some new pest that had arisen to plague the previous IR variety.[11]

Planting genetically identical strains across wide reaches of the landscape can also render crops throughout an entire region susceptible to disease outbreaks. The classic example has become the southern corn-leaf blight, a fungal disease that destroyed 15 percent of the U.S. corn crop in 1970 and cost American farmers $1 billion.[12] That loss was far from isolated. The genetic homogenization of tobacco strains has been blamed for a mildew outbreak that wiped out 90 percent of the Cuban tobacco harvest in 1980.[13]

Today's crop breeders must work constantly to improve the resistance of their genetically uniform, high-yielding crop varieties, a never-ending process that often relies heavily on the genetic library of codes carried by our dwindling stocks of traditional cultivars and by the wild ancestors of crop species. Private groups and government seed banks around the world are pushing to collect and preserve heirloom varieties and wild relatives of crops, and a few organizations are trying to promote genetic diversity in agriculture. Yet plant resources continue to be lost at a rapid pace, as land is cleared for agriculture and for expanding human settlements. Ironically, these are the very forces that are helping to drive the genetic erosion of our food crops and make the preservation of diverse plant resources so vital.[14]

Multicrop Agriculture

Many factors, such as population growth, urbanization, consolidation of small holdings into larger farms, mechanization, and government subsidies, encourage farmers to simplify their planting schemes and switch to growing cash crops. Yet, in recent decades, concerns about the long-term sustainability of intensive monocultures and the environmental degradation caused by the chemicals and technologies needed to support them, have fueled a growing interest in various types of multicrop agriculture.

The question for researchers is whether multicrop systems involving only two or three species can achieve the same yield as monocultures without sacrificing the sustainability that diversity provides. Unlike the green revolution approach, there is no easy, one-size-fits-all solution to multicropping. Yet studies so far show that the right combination of species in the right setting can consistently "overyield"—that is, yield more together than the same species grown singly on the same plot. Not only do multicrop systems often match the output of monocultures; more important, they may provide steadier yields through good and bad years while requiring fewer artificial resources to defend or nurture them.[15]

Widely used strategies involve combining a cereal grain with a nitrogen-fixing legume, such as interplanting rows of corn between rows of clover. The clover supplies nitrogen, provides a living mulch, protects soil and water, and creates habitat for insects that prey on crop pests.[16] Another common pairing is millet and cowpea, which, when interplanted, can yield 50 percent more than either crop grown singly.[17] Yet another yield-boosting strategy calls for alternating rows of sorghum and pigeon pea. Grown together in one test, the sorghum yielded 95 percent of what it would have in monoculture, while the pigeon peas, which continued to grow after the grain harvest, yielded 72 percent of what they would have if planted alone. Thus the total crop yield was much higher overall than if only a single crop had been grown.[18]

Even so, high yield may not be as important to small farmers as stability of yield and income. Preliminary evidence shows that at least some crop mixtures, including sorghum and pigeon peas, can provide

*In India, sorghum and pigeon peas planted together provide a more stable yield
year to year than either crop planted singly.*

that stability. An analysis of ninety-four trials in India, where sorghum
and pigeon peas are traditionally planted together, showed that monocultures of either plant experience greater fluctuations in yield than the
two grown together. In fact, calculations show that pigeon pea planted
alone is likely to fail one year in five, sorghum alone one year in eight,
but a multicrop only one year in thirty-six.[19]

Plant diversity in a field not only lowers the risk of complete crop failure but also preserves a level of natural pest control that helps reduce
losses to insects. In the Philippines, for instance, fewer maize borers attack maize when it is multicropped with peanuts. In Nigeria, the range
of cowpea cultivation has been extended by planting cowpeas amid
mixtures of cereals such as sorghum and millet, which protect the cowpea from insect damage.[20] Planting blackberries among the grapes in
California vineyards helps to control leafhoppers, and a rich under-
growth of flowers helps to stem pest attacks in other fruit orchards. Just
how these tactics, which home gardeners know as companion planting,
work in any given case is largely unknown. The companion plants may
provide habitat and nourishment for the enemies of crop pests, release

aromatic compounds that repel pests, or camouflage some attractant signal from the target plant. Even the plant diversity provided by hedgerows, woodlots, windrows, fallows, and nearby natural areas can sustain beneficial insects that help to prevent crop losses.[21]

Plant diversity can also help sustain productivity by preventing soil deterioration. In the tropics, where bare soil can quickly be ruined by leaching and weathering, some of the most successful multicrops are agroforestry systems where trees and often perennial shrubs such as coffee, cocoa, and cassava are interplanted with annual food or forage crops. In the southern part of the Philippines, for instance, a successful agroforestry approach called the Sloping Agricultural Land Technology was developed for highly erodible lands. Double rows of nitrogen-fixing trees and shrubs such as cocoa, coffee, banana, and citrus are planted along the contours of the slope, alternating with bands of annual crops such as rice, maize, tomatoes, and beans. It was hoped that, by stabilizing the slopes and enriching and protecting the soil, a continuous harvest might be possible. The results have been dramatic: an almost sixtyfold reduction in soil erosion and higher yields that brought a seven-fold increase in income to the farmers over a decade.

Unfortunately, the system has not spread widely outside the test area for reasons that have nothing to do with its ecological soundness. Uncertain land tenure and heavy up-front labor costs often stymie the planting of perennial shrub and tree crops in the tropics when farmers can't be assured they will reap the long-term benefits.[22]

In any multicrop system, of course, boosting output depends on picking the right combination of species. Jack Ewel and his colleagues set up an experimental timber plantation at La Selva Biological Station in Costa Rica using three fast-growing native hardwoods. Each tree was planted either singly or interspersed with understory plants, including palms that produce edible buds and fruit and heliconias, which are flowering plants related to bananas. The resulting productivity varied according to the tree species. One, a massive evergreen known as pilon, used resources so effectively that it suppressed the growth of the understory plants and produced just as prolifically with or without them. The understory plants significantly enhanced total productivity when planted with either of the other tree species—laurel and Spanish cedar, the latter a valuable timber species related to mahogany.[23]

Thus, it appears that the key to both high productivity and sustainability is finding the right combination of species—plants that complement one another's use of resources and buffer the system from one another's failures. Ewel suggests that the diversity of form is more important in successful combinations than the number of species, per se. In other words, the largest payoff often comes from adding plants with new architectures or functional traits, such as tree crops, shrubs, melon vines, and legumes rather than simply adding more grain species to a cornfield. Researchers use terms such as "integration efficiency" or "resource augmentation" to describe the way certain species complement one another and improve the overall productivity of the system, although it has proven difficult to say just why this occurs. Presumably, adding a species that can capture light, water, or nutrients that were previously going unused or even leaking from the system offers the best chance of boosting productivity.[24]

Experimental Systems

In recent years, studies inside high-tech growth chambers in the United Kingdom and in field plots in the North American grasslands have begun to assess productivity in higher-diversity systems that more fully represent wild plant communities. In both of these types of trials, greater species richness has been found to produce a more prolific bounty of plant material.

One of the trials was conducted by Shahid Naeem, now of the University of Minnesota, and colleagues using a series of controlled-environment chambers inside the Ecotron facility at Imperial College near London. The ecologists assembled fourteen one-meter-square artificial ecosystems, each with an equal number of plants, but a different species count. The most species-rich systems had sixteen annuals, the medium-diversity systems five, and the low-diversity plots two. Each lower-diversity community was a subset of the richest one—a mimic of natural communities that have been impoverished by species losses.

After monitoring the chambers for a full growing season, the team found that the highest-diversity systems both consumed the most carbon dioxide and produced the most biomass. To test the validity of their results, the researchers randomly selected another 150 combinations of two, five, and eight species from their sixteen-species pool and grew these in a greenhouse. Although some species-rich mixes turned out to be rather poor producers, and some species-poor mixes were prolific, on average the productivity correlated with species number.

In the relatively fertile soils and cozy atmosphere of the Ecotron chambers, light seemed to be the limiting factor in plant growth and productivity. Naeem and his colleagues believe the high-diversity systems grew more lushly because they included a diversity of plant architectures, from tall herbs to creepers, that formed a dense, multilayered canopy and took advantage of nearly every speck of light for photosynthesis.[25]

A more extensive study of the relationship between biodiversity and productivity was undertaken by David Tilman of the University of Minnesota. On a sandplain in Minnesota's Cedar Creek Natural History Area, he and his team sectioned off 147 plots, each three-by-three meters, and seeded them with species drawn from a pool of twenty-four North American prairie plants. Each plot got a random draw of one, two, four, six, eight, twelve, or all twenty-four species of grasses and wildflowers. After two seasons of growth, the higher-diversity plots had significantly greater productivity and also used soil nitrogen more efficiently than the less diverse plots. Undisturbed plots of native grassland nearby showed similar results.

In contrast to the Ecotron communities, Tilman thinks productivity in the prairie plant assemblages was limited by meager soil nitrogen supplies rather than light. About one-half of the enhanced production of the high-diversity plots could probably be attributed to a few aggressive plant species: yarrow, black-eyed Susan, blue grama grass, bunchgrass, and clover. But the other half, Tilman believes, was attributable to greater variation in rooting patterns that permitted more efficient exploitation of soil nitrogen.[26]

Sustaining Productivity

The plant assemblages used in the Ecotron and grasslands studies were clearly more species-rich than most multicrop agricultural systems. But just how much diversity does it take to optimize production at a given site in a single season? In the small prairie plots in Minnesota, the positive effects of species richness leveled off after the first ten species, so on average, twenty-four were no more efficient and productive than ten. That's far fewer species than exist in even the most impoverished temperate or tropical ecosystems. Could the earth really sustain itself with such an abbreviated cast of producers?

Although the question cannot be answered unequivocally, recent findings certainly suggest that what might appear to be extravagant and unnecessary species abundance in a single season may be critical to stabilize plant productivity over time.

For practical reasons, the few direct experiments that have examined the link between the functional resilience of a system and its species diversity have been conducted in grasslands and other fast-growing, low-stature communities rather than in forests. One of strongest results to date came from the Cedar Creek grasslands in Minnesota.

Tilman and his colleagues got an unexpected opportunity to look at the effects of drought stress the year Minnesota experienced one of its worst droughts on record. At the time, they were studying the effects of nitrogen fertilization on vegetation already growing in four grassy fields. Although researchers had known for more than a century that enriching infertile soils with nitrogen allows a few aggressive species to grow prolifically, reduce diversity by outcompeting other plants, and, ironically, boost biomass production, they did not know how well the productivity would persist. With the advent of drought, Tilman had a sudden opportunity to test productivity in the face of environmental stress.

The results were striking. During the dry years, productivity in all plots fell drastically. Yet in the most species-rich plots, productivity levels dropped only a fourth as much as in the species-poor plots. When the drought broke, production in the highest-diversity plots recovered in a single season, whereas the low-diversity plots needed four seasons to rebound to previous levels of productivity.[27]

One researcher, Tom Givnish of the University of Wisconsin, thinks the poor performance of the low-diversity plots may not reflect low species number per se, but the way those communities were impoverished in the first place. That is, heavy nitrogen fertilization favored the dominance of a fast-growing and prolific set of plant species that also required plenty of water and so were hard hit by the drought.[28]

Whether fertilization did indeed favor less stress-tolerant plants remains unanswered, but the issue is an important one, thanks in part to the unprecedented rain of nitrogen that humans spew into the air from automobiles, industry, and over-fertilized farm soils. In formerly nutrient-poor plant communities across western Europe, the increase in nitrogen deposited as windborne dust and acid rain is believed responsible for the large losses in biodiversity that have occurred in recent decades.[29]

For example, deposits of ammonium released by the heavy manuring of fields and pastures in The Netherlands is thought to be responsible for the displacement of once-dominant dwarf evergreen shrubs in the heathlands by a perennial grass. These grass-dominated communities are more than twice as productive as the shrubs, drive a much faster nutrient turnover, and also generate greater nutrient losses.[30] The question for the long term is whether this simplified community will withstand drought and other stresses as well as the one it replaced. Tilman notes, "The effects of nitrogen addition in lowering diversity per se are reason enough for concern about the stability of these communities." The possibility that nitrogen may at the same time be favoring a biased subset of plants with diminished staying power should spur greater concern about the impacts of profligate nitrogen emissions.

Stability through Compensation

After eleven years of monitoring, Tilman's fertilized plots in the Minnesota grasslands are now providing evidence of the mechanism by which diversity stabilizes productivity over time. Throughout the decade-plus, Tilman found that biomass production remained steadier in plots with more species of grasses and wildflowers. All the while, however, Tilman was also painstakingly charting the fate of each individual species, measuring its biomass in each plot in late summer.

He found that the productivity of individual species in high-diversity plots fluctuated considerably more than in the low-diversity ones. Why? Tilman suggests it is because competing species in a high-diversity plot were able to take advantage of one another's bad years, with some expanding to grab resources freed by others that were stressed by drought, cold, or other unfavorable working conditions. It turns out that this ability to compensate is what made the overall ecosystem more robust and kept production steadier over time.[31]

A handful of other experiments also support the notion that maintaining species richness among plants will help assure a consistently lusher world. Just as in the Minnesota study, these experiments show that diversity proves that some species fare better than others in times of stress, allowing them to take up the workload and keep ecosystem-level services, such as net primary productivity, fairly constant.

As early as 1954, German experimenter Heinz Ellenberg of Gottingen University showed that in a plot with four grass species, the abundance of each one rose or fell through wet and dry years while productivity of the whole assemblage remained rather steady. Two deep-rooted grasses grew well in a dry year when the water table in the sandy soils was low, yet fared poorly when the soil was waterlogged. In contrast, another of the grasses played a subordinate role in the dry years but performed so well in the wet year that without it, total community productivity would have dropped dramatically.[32]

More recently, in the upland tussock and wet meadow tundra of arctic Alaska, F. Stuart Chapin of the University of California, Berkeley, and Gaius Shaver of the Marine Biological Laboratory at Woods Hole found that no single environmental factor such as cold temperature, low light intensity, or low nutrient levels could inhibit the growth of all tundra species at once. Instead, compensation prevailed here too. When Chapin and Shaver altered the conditions in a patch of tundra by applying fertilizer or shadecloth or by heating it with a plastic greenhouse, some species declined but others grew robustly to compensate. With each manipulation, the contribution of each species of cotton grass, sedge, birch, or evergreen shifted dramatically. Overall, the diversity of their responses helped stabilize the system and keep total production fairly stable.[33]

In the serpentine grasslands of California, which are characterized by rocky, infertile soils, drastic year-to-year variations in rainfall combine with continual soil disturbance by gophers to keep the plant community in turmoil and enforce an ever-shifting patchy distribution of annual wildflowers and perennial grasses. When the rainfall in the spring is right, mounds of soil newly turned by gophers are quickly covered by aggressive colonizers like rosin weed, plantain, and creeping lotus, while yellow-flowering goldfields or small blue *Brodiaea* lilies thrive in the undisturbed soil nearby. Watching these dynamics for more than a decade in the Jasper Ridge Biological Preserve, Richard Hobbs of Australia's CSIRO Division of Wildlife and Ecology and Harold Mooney of Stanford have found that the continual disturbances prevent a few plant species from gaining dominance and so keep the community quite diverse. This results in higher and more constant productivity.[34]

Finally, on Tanzania's Serengeti Plain, grasses are subject to heavy foraging by some of the world's most spectacular nomadic grazers—about 3 million individuals of twenty-seven species, including wildebeest, zebra, Thomson's gazelle, buffalo, and topi. Samuel McNaughton of Syracuse University examined the impact of these grazers on grassland productivity over a number of years. He found that diversity was not linked to the most prolific productivity but rather to the most constant productivity. Specifically, good rains and moderate levels of grazing (rather than high plant diversity) produced the lushest grass crop. Nonetheless, the more diverse the plant community, the better it resisted losses to grazers, partly because the species-rich areas included a wide array of plants that certain grazers found unpalatable and therefore avoided eating. The species-rich grasslands also showed greater resilience, rebounding from the effects of grazing and recovering to a full standing crop more quickly at the onset of the rainy season.[35,36]

Plants for All Places

With so few studies in hand in such a limited range of ecosystems—mostly grasslands and tundra—researchers have also turned to indirect evidence to better understand the link between biodiversity and productivity.

Why, for instance, does such a seemingly extravagant abundance of plant species exist? The answer is in part because the earth's array of habitats are so various and complex, they provide opportunities for a myriad of specialists to develop. Even within a single biome, such as temperate forests or grasslands, there are numerous distinctive habitats and multiple types of ecosystems. Naturalists long ago observed that certain plants grow only on north-facing and not south-facing slopes, in dry uplands rather than moist lowlands, on certain soil types, at particular elevations, within specific temperature or rainfall ranges, and in the presence or absence of wind.

Even within a species, different genetic types may be keenly adapted to thrive best in certain portions of the species' territorial range. Populations of Douglas firs in the northern Rocky Mountain region, for example, are finely tuned for environmental conditions found only at certain elevations. Seedlings from lower elevations are known to be vulnerable to early fall frosts when planted at higher elevations. In contrast, seedlings from the higher subalpine zone show various adaptations for frost hardiness. Because of such findings, forestry researchers recommend that seeds used for reforestation projects in northern Idaho and eastern Washington not be moved more than 140 meters in elevation, nor more than 1.6 degrees latitude or 2.7 degrees longitude from their source. One Douglas fir, it turns out, is not the same as any other when it comes to maximizing growth and productivity under all conditions.[37]

Indeed, evolution seems to have produced at least one specialist to fill even the most inhospitable sites on the planet. Variations in growth rates, nutrient uptake, photosynthetic and respiration rates, and other physiological traits give certain plants a survival edge on sites too dry, shady, salty, cold, windy, or infertile for their neighbors. Plants that grow rapidly and prolifically, set seed, and die in a single season may be optimal for exploiting some disturbed or fertile sites; long-lived plants that produce new leaves slowly may persist better on harsher sites.[38] The collective outcome of these individual adaptive strategies is a certain level of constancy in the ecological services to which each species unwittingly contributes.

Despite all that's just been said about varied sites and specialists, there are rare cases where a single generalist species dominates the entire range of habitats and conditions in a complex landscape (this is, of course, not so rare in human-managed landscapes). A prime example is the domination of large areas of Hawaiian rain forest by a single tree, *Metrosideros polymorpha*. Its dominion ranges from sea level up to the subalpine zone, from wet forests to bogs and young lava flows. The dominance of *Metrosideros* has less to do with the tree's own talents than with the lack of challengers. Because Hawaii is one of the most isolated spots on the earth, separated from the nearest continent by at least thirty-five hundred kilometers of ocean, only a few immigrants prior to the arrival of humans reached its shores—perhaps only one every twenty thousand years.[39] The island forests are now paying a heavy price for their exceptional simplification, however. Their low diversity seems to make them especially vulnerable to invasion by exotic plants. The dominance of *Metrosideros,* for instance, is being particularly challenged by the imported nitrogen-fixing evergreen, *Myrica faya,* which has proven so adept at colonizing new lava flows (see Chapter Five).[40]

Patterns of diversity reflect not only variations in physical habitat but also changes in the same site through time. In many communities, the mix of plants on a given site may depend on the successional stage of the recovering vegetation following a disturbance such as a flood or fire. Experiments show that specialists in pioneering disturbed sites often don't thrive against the highly competitive species that arrive in later successional stages. Thus, maintaining stable productivity for the long term probably requires a balance between early and late successional plant species, some of which are highly transient.[41]

How Many Producers Do We Need?

All of these observations have led researchers to conclude that a plant community fully adapted to local conditions, and thus maximally productive through time, must contain at least one and perhaps a whole suite of species and genetic types well suited to each major habitat at each successional stage. That raises a question asked earlier: Just how many species of plants does an ecosystem need to maintain maximal, stable productivity? What is the threshold beyond which adding, or preserving, species has no further effect on productivity?

The answer depends first on how many different habitats or zones created by topography, soils, climate, or disturbance patterns exist in each biome. (*Biome* is a term used to designate regionwide ecosystem types, such as deserts or tropical rain forests.) In the admitted absence of solid data, ecologists working on the Global Biodiversity Assessment came up with a "rough but reasonable estimate" that there exist "on the order of ten substantially different zones per biome, with a range of perhaps four to forty." The estimate doesn't mean, however, that ten major plant species are likely to be adequate in any given place, since some species might duplicate talents rather than adding traits that augment productivity. Because of this, the ecologists doubled the threshold number to twenty species, on average, before the benefits of biodiversity to productivity begin to level off. It should be obvious that this estimate carries an enormous uncertainty factor.[42]

Yet the number is not out of line with the few experimental findings scientists have so far, from the temperate forests of the world to the Minnesota grasslands. It also falls in line with analyses from some of the earth's most species-rich systems. Joseph Wright of the Smithsonian Tropical Research Institute in Panama, comparing the productivity of tropical forests with that of timber plantations, found little reason to believe that the full wealth of species in the undisturbed forest is needed for maximal productivity. Indeed, he thinks that productivity peaks "at levels of plant species richness well below those commonly observed in tropical forests."[43]

By these estimates, even the most depauperate tropical forests and other wildlands today probably retain more than enough plant species

to maintain maximal lushness, at least for the short term. But such estimates are based on the status quo and fail to take into account the potentially catastrophic changes wrought by worsening air pollution, ozone
depletion, global warming, and all the other human-caused and natural
upheavals to come.

Even ignoring these future stresses, however, there are unassailable
reasons to regard that hypothetical threshold of twenty species as an absolute minimum for maintaining productivity. The reason is that species
are interdependent. Trees, wildflowers, grasses, and shrubs cannot survive and produce for long without the services of the birds, bees, and
bats that pollinate them; the mammals and birds that disperse their
seeds; insect predators that control plant pests; grazers that help shape
the plant community; and, finally, the decomposers. Each of these organisms, in turn, depends on relationships with other plants and animals in a widening web of linkages that stretches far beyond the original twenty producers. In addition, if biodiversity were pared down
anywhere close to that hypothetical number, the losses of other links in
the community web would eventually reverberate to doom many of the
chosen twenty.

CHAPTER SEVEN

The Power to Shape the Land

*A*t the end of the last ice age on the windswept steppes of Beringia, three dramatic transformations took place. One was the arrival of modern human hunters who crossed the Bering Strait land bridge from Siberia to Alaska armed with the latest advances in Stone-Age spear points. Another was the conversion of the vast, dry steppe grasslands into the wet moss and shrub tundra that prevails today across the arctic reaches of Alaska and the Yukon. Third was the rapid disappearance of the great beasts known collectively as the Pleistocene megafauna: woolly mammoths, mastodons, camels, horses, bison, and other animals weighing a ton or more. For decades, scientists have debated the links between those three happenings some ten thousand to twelve thousand years ago. The central question is: What or who killed off the megafauna?

Although the answers remain speculative, they fall into three camps: a wetter climate, human overkill, or a newly developed scenario that combines human impacts with the power of the megafauna themselves to shape the landscape.

Some scientists believe the climate turned wetter as well as warmer toward the end of the Pleistocene as the last ice age waned. In this new climate, the theory goes, grasses lost ground to mosses and shrubs, which are unpalatable to most grazing animals, and this climate-driven change in the landscape eventually starved out the massive herbivores. Yet critics point out that previous glacial periods had come and gone with no attendant increase in extinctions. Others believe instead that an unprecedented blitzkrieg by newly arrived hunters slaughtered the great beasts down to the last mammoth. If that is the case, why did so many birds and smaller mammals that should have been less vulnerable to hunters disappear along with the megafauna?

The third scenario assigns the megafauna keystone status and suggests that once human hunting of these architects of the steppe began, the reverberations of their decline may have altered the landscape and doomed the rest of the community as well.

Counterintuitive as it may seem, grasslands on the cold, wet soils of the Far North are unstable without grazing animals. Grazing stimulates the regrowth of sweet young grass shoots, and dung from the grazers fertilizes the renewal. The new shoots transpire at a high rate, drying and oxygenating the soil. This improves the habitat for soil microbes and spurs them to decompose litter and dung more quickly, thus recirculating nutrients to grow more grass to feed more grazers. Because most of the ice age megafauna were grazers, scientists assume they played a key role in nurturing the grassy steppe.

Suppose then that human hunters, recently arrived across the land bridge, began to take a significant toll on mammoths, mastodons, and other ponderous herbivores. As the herds thinned, more patches of grass remained ungrazed. These grew more slowly and transpired less as the stalks aged, leaving the soil sodden. Dead grasses piled up faster than the soil microbes could process them. The thatch of dry grass then reflected away the sun's heat, cooling the now-soggy soil beneath it and promoting the growth of mosses, which are low in nitrogen and hard for animals to digest. The surviving grazers avoided patches of the landscape where these unpalatable plants prevailed, and the lack of their grazing services only furthered the decline of the grasses. Finally, the grassy steppe gave way to the shrubs, sedges, and lichens of the tundra, changing the terms of life for every animal, large or small. Whether the last mammoths and mastodons starved or fell prey to human hunters, the great beasts themselves unwittingly helped to alter the face of a continent.[1]

A landscape, to an ecologist, is a patchwork quilt of interacting ecosystems arranged across a geographic area. Scientists have been slow to acknowledge the influence of living organisms on the patterns of these "earthquilts," instead giving climate and geology virtually all of the credit for determining the layout of forests, grasslands, deserts, and other types of biological communities across the land. Yet the power to shape the landscape did not vanish with the mammoths. Most ecosystems today contain pivotal species of animals and even plants whose activities or traits can alter the character and function of the landscape.

In recent millennia, of course, our own species has played the most pivotal role in reshaping the earth by felling forests, plowing grasslands, filling wetlands, and transforming other natural landscapes to agricultural and urban uses. A recent study, in fact, indicates that we and our earth-moving machines would now win hands-down in one-on-one comparisons with glaciers, winds, tectonic uplift, or any other geological force in shifting the very rock and soil of the planet into new configurations. Geomorphologist Roger Hooke of the University of Minnesota calculates that humans worldwide move 40 billion tons of earth each year—churning it up during mining operations and construction and indirectly eroding it away by plowing croplands and clearcutting forests. That toll surpasses even the work of seafloor volcanoes, which thrust 30 billion tons of new material into the spreading seams of the earth's crust each year. Only rivers, surging and meandering through their floodplains, constantly remodeling their channels and banks, routinely rival humans with bulldozers for tonnage of earth moved.[2]

Certainly no other organism can make such grand claims as an earth shaper. There is a critical difference between our work and that of other creatures, though. Humans may sculpt the earth more dramatically, but we seldom make it more ecologically complex. In fact, our specialty is simplifying the landscape, turning diverse forests or meadows into tree farms, uniform rows of grain, or monotonous expanses of concrete and lawn. Human-managed lands seldom achieve—or ever aim for—the dynamic complexity inherent in natural systems. In contrast, many animals and plants routinely foster a more heterogeneous and dynamic landscape than the forces of climate, geology, or human activities could create.[3]

Throughout history, humans have worked, wittingly or not, to reduce or eliminate the influence of the landscape's other strong players, from beavers to prairie dogs to elephants. Land managers often proceed on the assumption that the physical disruptions provoked by influential organisms will thwart their efforts to hold timber or rangelands in the most economically desirable state. In part because of such concerns, human intervention has steadily diminished the variety and numbers of these organisms, as well as the land area they control.

Yet the loss of earth-shaping organisms, or their introduction to new settings, may trigger unwanted changes in the landscape. A world without elephants, moose, prairie dogs, mangrove swamps, and coral reefs would not just be an emptier version of the same scenery. Slowly, inexorably, the processes of earth, fire, and water would change, too, and so would the forests, plains, savannas, or coastlines.

Unfortunately, it has been easy to overlook these consequences of biodiversity loss. Often, influential organisms disappear as a side effect of land clearing or other direct human alterations of the landscape. When unwanted changes in ecological processes result, it can be difficult to track whether they were caused by the physical changes in the land or by the loss of biodiversity.[4]

In recent years, ecologists have begun to pay more attention to the impacts of organisms on the land and the kinds of changes that can occur when their work is disrupted. For instance, natural landscapes consist of a mosaic of habitat types, all of them linked by the movement of water, nutrients, energy, seeds, and animals. Changes in biodiversity that alter the distribution or flow of these resources can affect the landscape's workings, especially its productivity and the conservation of water, soil, and nutrients. Also, most landscapes are subject to periodic disturbances, such as fire, flood, or hurricanes. Losses or additions of organisms that influence the frequency or intensity of disturbance can strongly change the dynamics of the landscape. Most systems are also vulnerable to losses of key animal species whose daily activities modify the landscape, either directly, through physical changes like tunneling or toppling trees, or indirectly, through what they eat or do not eat.

Indeed, the events that took place in Beringia during the late Pleistocene were just one chapter in a larger story that suggests what might happen to today's landscapes and the biodiversity they nurture if the world's surviving megafauna go the way of the mammoths.

Modern Megaherbivores

Beringia was the easternmost extension of a vast, unglaciated steppe that stretched across northern Asia and into Europe at the end of the last Ice Age. South of Beringia, below the great Laurentide and Cordilleran ice sheets lay the ice-free plains of North America. As the climate warmed and the ice retreated, expanding human populations pressed north across Eurasia, into Beringia, and south from there throughout the Americas. During that period, virtually all the megafauna, not just of Beringia but of Europe and the Americas as well, went extinct. Furthermore, three-fourths of the smaller creatures, weighing one-tenth of a ton to just under a ton, also vanished. Losses throughout North and South America were particularly severe. About three-fourths of the animal genera living there at the time disappeared: mammoths, mastodons, proto-elephants called gomphotheres, hippo-like Toxodons, bear-sized beavers, bisons, camels, horses, giant ground sloths, armadillo-like glyptodons, and the cheetahs, lions, saber-toothed cats, and other predators and scavengers that lived off these colossal vegetarians. (See Chapter Three for more on the loss of large fruit-eaters, such as gomphotheres, from Central America.)

As in Beringia, it's plausible that human hunters could have eliminated the larger, slow-to-mature animals by killing more than were born each year. It's hard to imagine, though, that our forebears killed off the abundant, fast-reproducing smaller mammals. As has been proposed for Beringia, perhaps the loss of colossal grazers from the American grasslands changed the plant community so drastically that it made life untenable for most of the other animals, thus setting off a cascade of extinctions. Ecologist Norman Owen-Smith of the University of Witwatersrand in South Africa has called this the "keystone herbivore" hypothesis.

To get an idea of what might have happened on the North American plains ten thousand years ago, Owen-Smith focused on the role that the earth's surviving megafauna—elephants, rhinoceroses, and hippopotamuses—play today in shaping the African landscape. He has designated these animals keystone herbivores because their feeding habits can radically alter the composition and structure of the plant community across broad reaches of the savanna, dictating the living conditions for many other creatures.[5]

Elephants, for their part, are powerful forces for the restoration of grasslands. Grasses and woody plants coexist in a dynamic balance on the savanna. Overgrazing of the grasses brings on an encroachment of bush and thicket—a loss of economically useful range that humans often classify as desertification and attempt with little success to reverse with bulldozers, plows, and herbicides (see Chapter Four). Yet browsing elephants do the job effortlessly. As they feed, elephants break off trees and shrubs, uproot them, or push them over, often devouring all but the trunks and major limbs, trampling what they don't eat, and quickly turning dense woodlands back into open grasslands or shrubby regrowth. A single adult elephant may topple four or more trees each day, about fifteen hundred a year.[6] Only fire can rival the power of elephants in opening up the savanna.

White rhinos and hippos take up where elephants leave off, creating mosaics of diverse habitat in the tall grasslands opened up by elephants. With their grazing, they transform patches of tall grass into "grazing lawns" of more nutritious, less fibrous short grasses preferred by pickier grazers, such as wildebeest, zebra, and impala. Because these closely cropped lawns accumulate too little dry grass to fuel wildfires that would otherwise kill off encroaching shrub and tree seedlings, rapidly regenerating thickets may reinvade, at least until elephants reappear.

This interplay of megaherbivores results in a patchwork savanna where a wide array of grazers and browsers can make a living off the productive and succulent short grasses and young shrubs, while those who require it can still find cover from predators in patches of tall grass or thicket.

Imagine now the interplay of megaherbivores in North America south of the ice sheets ten thousand years ago. The pollen record indicates that a wide sweep of the continent from the Appalachians to the Rockies was covered by open, parklike woodlands where conifer and hardwood trees were interspersed with stretches of grass and wildflowers. In this setting, browsing mastodons, gomphotheres, and ground sloths might have filled the elephant role while grazing mammoths took the place of white rhinos and hippos. At least until humans arrived, the adult megaherbivores were probably fairly immune to predators, and so overall numbers could have been high, in fact "vastly greater than anything yet documented in Africa," Owen-Smith speculates.

Then, in an astonishingly short time, the megaherbivores vanished. Not long thereafter, open glades would have filled in with unbrowsed shrubs and seedlings, crowding out the grasses and herbs. The ungrazed grasses in the clearings would have grown tall, fueling more intense and frequent fires that killed off the seedlings in their midst. The result, Owen-Smith suggests, was the conversion of those mixed savannas into distinct zones of dense forest and uniform prairies typical of the region today. Without megaherbivores to create and maintain a diverse patchwork of habitats, the populations of many medium-sized and smaller creatures would have been fragmented into isolated and shrinking pockets of suitable living space. With escape routes increasingly blocked by open prairie or dense forests, the small mammals would have been at the mercy of a shifting climate, chance disturbances, and even human hunters. Eventually most of these creatures went extinct, too.[7]

Today, the influence of megaherbivores remains largely confined to the game parks of Africa, where managers find it increasingly difficult to balance the creative and destructive powers of these beasts. No park, and certainly no other African ecosystem today, contains a full complement of both browsing and grazing megaherbivores: white rhinos, hippos, and elephants. Outside the parks, few regions remain sparsely peopled enough to allow elephants to practice the hit-and-run remodeling of the savanna that once kept them ranging over vast territories while the bush recovered in their wake.[8]

Over the past thirty years, as pressures from ivory poachers and a burgeoning human population have concentrated elephants into protected areas, many of these parks and reserves have experienced a decline in woodlands. The shift has been a subject of great concern to ecologists and park managers who remember the dense thickets common in savanna parks during the first half of this century. Yet the classic safari-brochure image of the savanna may have been an anomaly. For thousands of years before that, humans and fire and elephants had apparently maintained savanna Africa in a grassland state. Then, in the nineteenth century, hunters decimated elephant and rhino populations, and during the 1890s, rinderpest devastated large grazers such as buffalo, wildebeest, and domestic cattle. Pastoral people, who set most savanna fires to create grazing lawns for their cattle, largely starved or moved off. Bush and thicket took hold across wide stretches of savanna, and with it came tsetse flies and sleeping sickness. The bush and flies and disease discouraged human resettlement, and because these areas were largely unsettled during the colonial era, many were made into parks. Only when elephants began crowding into these parks in high numbers during the 1960s did the dense bush country begin reverting to grassy steppe.[9]

Some still debate whether elephants or fire bear the greatest responsibility for the recent growth of grasslands. Take the plains of Tanzania's Serengeti National Park and the adjoining Masai Mara Reserve in Kenya, where dense *Acacia* savannas have reverted to open grasslands since the 1960s. At that time, the vast migratory herds of wildebeest and zebra still had not rebounded from the rinderpest pandemic, although cattle vaccination programs had begun to control the spread of the disease. Elephant numbers were growing, but slowly. High rainfall and the absence of heavy grazing allowed luxuriant grasses on the eastern plains to proliferate, providing plenty of fuel for fierce and frequent fires. Holly Dublin of the University of British Columbia and her colleagues believe that these fires were largely responsible for the initial reversion to grasslands. They base their interpretation in part on computer models that show that the elephant population present at the time could not have destroyed trees fast enough to open up the woodlands alone. On the other hand, even the most conservative rate of human-caused burning during the 1960s would have been enough to clear the bush.

Today, the situation has changed; grazing herds have rebounded to spectacular numbers, dry grass to fuel fires is sparse, and fire incidence has declined. Yet the woodlands show no sign of rebounding. What is keeping the savanna locked in a grassland state now? By the 1980s, the elephant population in the reserves had doubled, and Dublin and her team calculate that elephants are now a powerful enough force to prevent woodland encroachment. If park managers wanted woodlands in the Serengeti-Mara to regrow, the team found, they would have to reduce fires to an unrealistically low level and also eliminate more than half of the elephants in the parks.[10]

While culling of elephants is not done in these East African parks, managers of many parks in southern Africa regularly reduce elephant numbers to allow regrowth of wooded areas. Even in parks that are not culled, ecologists have long been concerned that unchecked elephant numbers will eventually destroy the very habitat mosaic that sustains both the elephants and many other savanna dwellers as well.

A worst-case preview of what could lie ahead occurred in Tsavo East National Park in Kenya after large herds of elephants took refuge from poachers and human settlers in a marginal area with low rainfall and proceeded to destroy the bush thickets. During a severe drought in 1970–1971, a third of the park's elephants, between six thousand and ten thousand animals, died of starvation.[11,12]

The effects of intensive elephant activity on the well-being of other species is harder to judge. At Tsavo, the decline of bush habitat reduced populations of browsers such as black rhinos and lesser kudu and increased populations of grazers such as Grevy's zebra and oryx. In nearby Amboseli National Park, bushbuck, lesser kudu, giraffe, and vervet monkeys present in the 1960s have disappeared, along with half of the park's plant species and most of its woodlands, the latter ripped and trampled to a moonscape by unprecedented crowds of elephants.[13] Yet, in other parks, perhaps due to differing soils and rainfall, the opening up of dense thickets has benefited browsers such as kudu and eland by making forage more accessible.

This Solomonic dilemma that many park managers find themselves facing today as they try to balance the needs of woodland and grassland species in a constricted area tragically demonstrates the unprecedented loss of wildlands that has marked the twentieth century. The same quandary faces managers who must balance the needs of creatures that require old-growth versus pioneering forests, wet meadows versus open-water marsh, and many other seemingly dichotomous habitats that once persisted in concert across vast quiltwork landscapes.

While the problems created by elephants confined at high densities are very real in many parks, from Amboseli in Kenya to Chobe in northern Botswana, the long-term threat for African savanna communities is actually the opposite: the extinction of elephants and the other surviving megaherbivores. What would the loss of Africa's megaherbivores mean for the landscape and the other mammals and birds of the savannas? Owen-Smith is hopeful that both erratic rainfall and geological variations would keep the African savanna more diverse and patchy than the plains of North America, even if the megaherbivores were lost. However, the history of South Africa's Hluhlue Game Reserve suggests that in some African ecosystems, the loss of elephants, rhinos, and hippos could trigger an inexorable decline among large grazers and eventual loss of other creatures as well.

In Hluhlue and much of the rest of Zululand, nineteenth-century ivory hunters had eliminated all elephants by 1890. Bush invasion followed, and since the 1930s, reserve managers have used fire, chemical herbicides, and manual clearing in a continual attempt to halt the encroachment of wood and thicket. Theirs has been a losing battle. With the advance of the bush, three species of antelope, including reedbuck, have gone locally extinct, and open-country grazers, such as wildebeest, have declined. Even some browsers, such as black rhino, have dwindled. In 1981, the park reintroduced elephants in hopes that they may be more effective than humans at restoring the grasslands. By 1989 Hluhlue had one hundred elephants, but it may be decades before this experiment shows results.

Unfortunately, conservation campaigns for elephants and rhinos often portray them as mere symbols, charismatic standard-bearers for their communities at best, or sentimental anachronisms at worst. Yet the survival of the savanna community is inextricably bound to their fate. If these pivotal architects of the savanna are lost, attempts to preserve the full richness of large mammals that has persisted in this one last strong-hold since the Pleistocene may ultimately fail.[14]

Four-Legged Architects, Engineers, and Gardeners

While no continent besides Africa came through the ice ages with its megafauna intact, the rest still harbor smaller mammals that sculpt and diversify the landscape through their daily activities. Most work on a lesser scale, or more slowly or subtly than elephants, but all influence the kinds of plants that persist, the successional path the plant community will follow, and the richness of biological activity in a landscape. Although their influence has been diminished as humans have usurped control of the land and reduced their populations, beavers, moose, prairie dogs, pocket gophers, lemmings, grizzly bears, and many others still make their mark on the landscape.

The most ubiquitous and well-known earth-shapers are beavers, whose engineering works once influenced the course of nearly every wooded stream or lake throughout Europe, Asia, and the Americas. Since hitting a nadir at the turn of the century, thanks largely to fur trapping, beaver populations have been rebounding naturally or have been reintroduced across much of their historic range.

The nesting, burrowing, and food caching activities of beavers are similar to those of other rodents, except they take place in the water and therefore alter the hydrology and most biological processes in the stream. These activities require a number of traits unique to beavers: the ability to cut down trees, build dams, and impound flowing water. As they gnaw through tree trunks with their powerful incisors, drag trunks and branches into the stream to build dams and food caches, and build lodges in the open water of their ponds, beavers also open up gaps in the forest canopy and establish a patchwork of habitats—moist meadows, wetlands, bogs, and ponds—used by many other creatures from waterfowl to spawning fish. (See also Chapter Four.)[15]

In New York's Adirondack State Park, researchers found that abandoned and drained beaver ponds are usually reclaimed and reflooded after ten to thirty years. Marguerite Remillard of the University of Georgia and her colleagues used aerial photographs to follow changes in vegetation at thirty-nine relatively isolated beaver colonies over a forty-year period. When beaver dams were breached and the ponds drained, the clearings that resulted did not show a predictable pattern of plant succession in their enriched, nutrient-laden sediments. Many became so-called beaver meadows; in some of them, shrubs like alder later sprang up. Eventually these might have given way to hardwood and conifer saplings, as happens when beaver are completely eliminated from a region, but the researchers found that this progression was likely to be interrupted and the site reoccupied. In fact, none of the Adirondack sites got beyond the shrub stage before being converted to open water again by returning beavers. Ponds typically remained flooded and active for eight to ten years at a stretch, usually until the beavers depleted the supply of preferred trees in the nearby woods. Then after ten to thirty years of abandonment, returning beavers would begin the cycle again.[16]

Aspen, willow, and other hardwoods are the beavers' favored trees for food and building materials, and the animals usually cut large-diameter trees—those with trunks at least ten centimeters across. In a classic northern forest succession, trees like aspen and willow normally spring up after fires and blowdowns and then are overtaken and eventually shaded out by conifers, such as spruce. Yet when beavers assume control of a site, the regrowth of the forest around their pond suddenly changes. The opening of the forest canopy causes hardwood stumps and roots to sprout prolifically, generating dense stands of suckers and stump-sprouted hardwoods around the pond. This process stalls the normal successional pattern, allowing the hardwoods to persist longer than they otherwise would.

The influence of beavers on northern forests is countered in some places by the voracious and selective appetite of another key animal architect, the moose. Like beavers, moose forage on tender and palatable hardwoods, although they select the smaller diameter trees that beavers usually pass up. In fact, moose often browse on the hardwood regrowth around beaver ponds, where their heavy feeding—five to six tons per year—can limit hardwood growth enough to allow spruce, balsam fir, and other conifers to take over.

Significantly, the moose avoid eating spruce and other conifers. That's because the conifers carry a heavier load of lignin and resin and supply less nitrogen than hardwoods. Neither the microbes in a moose's gut nor those on the forest floor find it easy to digest conifer wood and needles. By avoiding spruce and suppressing young hardwoods, however, moose inadvertently speed the march toward spruce dominance and, ultimately, a boom-and-bust oscillation in population dynamics in this ecosystem. The moose also set in motion a significant change in nitrogen cycling, suppressing the hardwood trees that supply easily rotted, high-nitrogen leaf litter and thus favoring those that carpet the forest floor with slow-rotting needles. When John Pastor of the University of Minnesota and his colleagues tested the soils at various sites on Isle Royale, Michigan, they found that the amount of nitrogen available to plants was highest in areas where beaver cutting encouraged hardwood sprouting and lowest where moose browse heavily. Since hardwoods require more nitrogen for growth than conifers, the decline in soil fertility in areas where moose browse reinforces the advance of the spruce.

Spruce cannot dominate the boreal forest for long. Their resinous tissues are more flammable than hardwoods, so the more spruce there are, the greater the probability of forest fire. Eventually, fire, high winds, attacks of spruce budworm, or even a diminished nitrogen supply will destroy large areas of spruce, allowing hardwoods to recolonize in the nutrient-rich ash and thus restarting the successional cycle.[17,18]

A moose browsing on willow twigs may not convey the same sense of power over the fate of the land as an elephant pushing over trees or a beaver gnawing them down. Yet the changes set in motion by the moose can just as surely, if more slowly, drive the fate of its forest. Paradoxically, as the forest goes, so go the moose, and these creatures seem by human standards to be working against themselves. Although moose shelter in the spruce, they cannot eat them, so the end result of the moose's labors is eventually a drastic crash in its own populations. Other boreal and tundra species, such as lemmings, voles, wolves, hares, and lynx are legendary for their wild swings in population too. Pastor and his colleagues believe these little-understood boom-and-bust oscillations are part of a unique dynamic that makes northern ecosystems qualitatively different from any other ecosystem on earth. If so, then the researchers believe that attempts to damp these swings by stabilizing moose numbers or suppressing fires will interfere with the integrity and diversity of these ecosystems.[19]

In many areas where the Pleistocene megafauna once roamed, small rodents have assumed the role of key grazers. From Russia to Alaska, brown lemmings graze in isolated pockets of productive and stable tundra grasslands. Such pockets can be found in ravines on the Russian tundra or troughs between strange formations called ice-wedge polygons on the Alaskan tundra. In troughs and gullies, lemmings can shelter under deep snow cover and stimulate grass growth by cropping it and fertilizing it with their droppings. During periodic population outbreaks,

Prairie dogs dig extensive burrows and clip the grass around them, creating shelter and nutritious forage for many other animals.

the lemmings expand their range, helping to create and maintain productive tundra meadows.[20]

Further south, in the grasslands of North America, a type of ground squirrel known as the prairie dog, *Cynomys ludovicianus,* disturbs and diversifies the landscape with its grazing and burrowing, creating a complex and biologically active system. Living in large colonies known as towns, the prairie dogs create areas of high soil fertility and nitrogen-rich short grasses that may span tens to hundreds of hectares across. No one knows how much land these rodents controlled prior to European settlement, but as of 1919, prairie dog colonies still covered some 40 million hectares, more than 20 percent of the short-grass prairie landscape in the United States. Today 98 percent of those populations have been eradicated in an ongoing control effort by range managers who view prairie dogs as pests that reduce the amount of grass available to cattle.

The ecological services provided by prairie dogs are twofold. The short-grass areas they help create provide food or shelter for many other animals, from pronghorn antelope and bison to mice and burrowing owls. In addition, the defensive encampments they build attract a plethora of predators from hawks, coyotes, snakes, badgers, and bobcats to black-footed ferrets. (The ferret is a highly endangered species that has spurred belated interest in conservation of prairie dogs because they are its only food source.) To evade these predators, prairie dogs dig extensive underground burrow systems with multiple entrances, piling and maintaining mounds of bare soil around the holes to serve as lookout posts. They also repeatedly clip the grass and wildflowers to half-height, apparently to make approaching predators more visible.

As a consequence, the grass in a prairie dog colony consists of a greater percentage of highly nutritious and digestible young shoots than does the surrounding prairie. This high-quality forage attracts native grazers such as bison, elk, and pronghorn antelope. Bison that feed in prairie dog towns regularly put on more weight than those that feed on the nearby prairie. There is still a fierce and ongoing debate between researchers and ranchers, however, over how cattle are affected by the tradeoff between higher quality but lower quantity of forage. The prevailing sentiment among ranchers in the western United States is that more prairie dogs mean fewer cattle on the range, and so eradication programs continue in most areas.[21]

Some researchers believe that the work of prairie dogs, gophers, squirrels, and other rodents that dig or burrow may be one of the major forces for churning and translocating soil and nutrients in grasslands. At least historically, their impact has been significant: early in this century, such rodents represented nearly one-fourth of the mammalian species, and perhaps one-half of all mammals in the western United States.[22]

One of these rodents, the pocket gopher *Geomys bursarius,* still ranges across the grasslands and scrublands of most of the western half of North America, and similar rodents can be found on other continents. Like prairie dogs, gophers can profoundly alter the topography of the landscape, the mix of plant species, and the behavior and number of their fellow grazers. Gophers do most of their feeding from their underground tunnels, eating the roots of grasses and forbs and sometimes young trees, and often pulling whole plants down through the soil into their tunnels. Even more than prairie dogs, these rodents are considered agricultural pests and are actively exterminated by farmers and gardeners in most regions.

The most obvious and long-lasting gopher signs on the landscape are their earthworks, known as mima mounds. In engineering their tunnel systems, gophers move dirt to the surface and mound it up just as prairie dogs do. The creatures' small size belies their power as earth-movers, and they seldom work in isolation. In fact, from fifty to one hundred gophers—sometimes more than two hundred—can be found per hectare. By one estimate, gophers in Yosemite National Park in California deposit at least eight thousand tons of soil on the surface every year. The result may be one hundred or more mima mounds per hectare, each up to two meters high and twenty-five to fifty meters across. Many western and midwestern grasslands display an undulating terrain created by thousands of years of gopher activity. Similar mima-type terrain from Argentina to the Cape of South Africa has also been attributed to tunneling rodents.

Despite the pesty reputation of gophers, the long-term effect of their activities is to create a patchier distribution of soil nitrogen that allows a more diverse collection of plants to coexist and often leads to greater plant productivity and diversity. It is even possible that gophers increase the abundance of their favorite food plants, "effectively farming their preferred resources."[23]

Rodents are not the only mammals that churn and till the soil, or appear to farm their choicest forage plants. On alpine meadows in the northern Rockies, foraging grizzly bears make their own mark on the land and the community, effectively reinforcing the growth and nutrient content of one of their favorite foods and also indirectly increasing the fertility of the soil.

Hikers in Glacier National Park on the U.S.–Canadian border often see large, dense patches of yellow glacier lilies on high meadows after the snowmelt, but few recognize them as the work of grizzly bears. Bears have a special fondness for the underground bulbs or corms from which glacier lilies arise. In the late summer and fall, grizzlies forage for these corms by clawing up and turning over chunks of sod—not a trivial task given the highly tangled root mass that holds the alpine turf in place. As they work, the bears aerate and warm the soil, speeding up the labors of microbial decomposers and allowing rainwater to soak in more easily. Sandra Tardiff and Jack Stanford at the University of Montana found that soils in these bear gardens, which may be fifteen meters square, contain two to five times as much nitrogen as the undisturbed meadow nearby. The lilies in these digs have higher nitrogen concentrations in their tissues, too, and produce more seeds. Because the lilies are good at colonizing disturbed soil, they sprout more readily in these tilled plots than in the surrounding meadow amid the woodrush, sedge, and miner's lettuce. Tardiff has observed that bears tend to return to these former digs, probably because the digging is easier and the corms are richer. As they feast, the bears assure a dense regrowth of the enriched lilies for the next season.[24]

Plants That Promote Fire

Fire is a major source of disturbance and renewal that profoundly shapes many landscapes, resetting the successional cycle and causing some fire-adapted plants to release seeds or germinate. As described above, certain types of vegetation in a community may greatly increase the chance of a burn. Mature spruce forests with their thick carpets of dry needles and trees loaded with flammable resins and dead branches are more combustible than hardwood stands and also burn hotter. Animals as well as plants are an influential part of this cycle. Not only do the appetites of moose and beavers help determine when a northern forest will reach this flammable spruce stage, but beavers also indirectly affect the extent of the burn. The mosaic of aspen and willow stands, meadows, ponds, and wetlands they maintain amid the flammable spruce forests can serve as natural firebreaks, keeping fires smaller than they would be in homogeneous landscapes. Both the moose and beaver then benefit from the effects of fire because it clears the way for the regrowth of aspen and willows in the nutrient-laden ash.

Many other plants exercise a controlling influence on the landscape by altering the probability of fire. Mistletoe, a parasitic plant that commonly infests pines, increases the flammability of a forest, apparently to the benefit of the parasite itself. It does so by encouraging high fuel loads. Infected trees often sport abnormal growths, such as resin-filled burls or finely branched "witches brooms" that are made more flammable by the piles of needles and other litter they trap. In managed forests, fire suppression efforts may actually enhance the spread of mistletoe and its effects, setting the stage for more intense and larger burns. These burns in turn may create conditions favorable for the proliferation of the parasite's host tree. After a mature forest burns, an abundance of recolonizing pines may spring up, providing more hosts for mistletoe and so perpetuating the cycle.[25]

The most dramatic demonstration of the influence of plants comes when fire-adapted alien species invade natural ecosystems. One such invasion began in the 1880s when Eurasian cheatgrass arrived in the Great Basin of North America, an arid, mostly treeless sagebrush steppe that stretches between the Rocky Mountains on the east and the Sierra Nevada and Cascade ranges on the west. Cheatgrass did not just slip into the community quietly. It utterly transformed the landscape. By speeding and intensifying the range fire cycle, cheatgrass has converted about 40 million hectares of diverse sagebrush community to a monotonous annual grassland that permits more soil erosion and reduces wildlife habitat.

Until the arrival of cheatgrass, some scientists think range fires must have been fairly rare, relatively cool, and small because fires do not carry well between sagebrush and bunchgrass clumps. Annual cheatgrass, however, produces several times more fuel than the native plants, encouraging hotter, more frequent wildfires that blaze quickly through continuous carpets of dried grass. Cheat's real talents show up after fires, however. Shrubs and perennial grasses are burned back to the roots or killed outright in hot fires. In contrast, cheatgrass sets copious seed before the summer fire season begins, and those seeds germinate in the fall and winter rains after the fires, creating a lush new grass cover that blooms and sets seed again the next spring, then dries to fuel another summer of fires. [26,27]

In recent years, cheatgrass has even begun to threaten large areas of conifer woodlands that are unique to the mountain slopes of the Great Basin. These are the pinyon pine and juniper stands that dominate the zone just above the sagebrush steppe. Cheatgrass has spread to the edges of this zone and now frequently carries fires into the woodland, too, destroying large areas of bird and wildlife habitat in this unique ecosystem.[28]

Cattlemen, for their part, remain ambivalent about cheatgrass. For six to eight weeks in the spring, cheat is lush and inviting forage for cattle. Later, however, it dries and turns almost useless. In fact, its barbed awns can blind cattle or cause mouth sores. In spite of this, reducing grazing to avoid range damage and help native vegetation resist the cheat invasion remains a hard sell among ranchers. A greater concern that ranchers cannot ignore, however, is the fact that this impoverished range community is ripe for invasion and further degradation by a host of even more noxious Eurasian weeds—medusahead rye, rush skeletonweed, musk thistle, spotted knapweed, leafy spurge, and dalmation toadflax—for which cattle have even less use.[29]

Further west, the native grasslands, chaparral, and coastal sage scrub communities in California also have been overrun by introduced grasses and other annuals over the past two centuries. In fact, many of the introductions into the chaparral have been deliberate attempts to transform the area into a more economically useful range. In recent decades, land managers have also sown exotic annual grasses into burned chaparral after brushfires in an effort to stabilize slopes before winter rains cause mudslides and floods. Unhappily, this quick fix often causes profound changes in the scrub community, which in its healthy state is more effective than the annual grasses at holding back erosion. The grass alters the scrub by encouraging another burn years before the natural vegetation would have produced enough biomass to fuel a second fire.

At one site on San Diego's Otay Mountain, for instance, forestry officials planted annual ryegrass amid a charred sage scrub and chaparral community following a two thousand-hectare arson fire in 1979. Heavy rains that winter produced a lush growth of ryegrass, but, unfortunately, by summer the dry grass provided a much bigger stock of flammable material than would have been produced by the sparser growth of native herbs and wildflowers that resprout after chaparral fires. That summer, another arson-set fire—a growing problem in California urban areas—caused a hot and extensive reburn of much of the charred slope.

When Paul Zedler and his colleagues at San Diego State University examined sections of the slope that had burned only in 1979, they found that mature shrubs like ceanothus and chamise had been killed, but their seeds and roots had resprouted prolifically after the fire. In fact, the regrowing shrubs had come back at a much higher density than before the fire. In the twice-burned areas, the second fire destroyed the shrub seedlings before they could bear seeds, so little was left to resprout. Ceanothus, chamise, and many other shrubs declined drastically, creating a conspicuously different plant community than on the once-burned slope. Zedler speculates that the ryegrass-dominated slope may continue to burn more frequently than the natural plant community nearby, preventing a return of the chaparral. The end result is an ecosystem with its structure and functions cut loose from natural successional cycles and driven, instead, by disturbance.[30]

Fire-enhancing weeds, especially grasses from Africa and North America, are also creating large changes in Hawaiian ecosystems where fire was once a rare occurrence. Unlike aliens in continental ecosystems, which often require a disturbance such as plowing or overgrazing to get started, some invading grasses apparently need little help to gain a beachhead in these isolated island communities. Alien grasses now dominate large swaths of the dry and seasonally dry habitats in Hawaii, and they carry fire from these areas into native communities that have never adapted to fire.

Before the late 1960s, for example, the woodlands on the southwest flank of Kilauea Volcano in Hawaii Volcanoes National Park were dominated by open expanses of ohia trees and understory shrubs with few native grasses. During nearly a half-century up to that time only nine fires were recorded in the woodlands on the lower slopes, and altogether they burned less than two and a half hectares. Then a dense carpet of alien grasses invaded the woodlands. In the nineteen years between 1969 and 1988, some thirty-two fires charred more than seventy-eight hundred hectares of the woods during the long annual dry seasons. The burns not only reduced the diversity and distribution of native plants but opened the way for expansion of alien grasses, which snap back and grow vigorously after fires. Eighteen years after the last burn, these grasses still dominate.[31]

Elsewhere in the tropics, one of the most serious weeds on rubber and other tree plantations is a vine-like perennial shrub called *Chromolaena odorata,* which not only infests cleared land but can invade nearby natural areas and carry wildfires into the tropical forests. Farmers from the Philippines to West Africa to the Indian subcontinent sometimes lose their battles to keep pastures and plantations clear of *Chromolaena.* The shrubs, which can grow to seven meters in height, produce massive quantities of wind-dispersed seeds that spring up to crowd out native vegetation. The mature shrubs also provide dense masses of dry biomass that burn easily. The species name *odorata* gives a clue to the large stock of essential oils, alkaloids, and other secondary chemicals this weed produces to protect it from grazing insects and other animals. The oils make the dry biomass highly flammable. In Indonesia and Malaysia, another agricultural weed, the tall grass known as cogon or alang-alang, carries fire into natural areas. In the dry forests of Costa Rica, tall jaragua grass performs the same function.[32]

In south Florida *Melaleuca quinquenervia,* an Australian tree that was introduced early in the century to help drain the swamps, is considered "the most serious threat to the integrity of all south Florida's natural systems," including the Everglades. That's no small feat in a state that harbors more successful invaders than any other in the continental United States.[33]

The secret of *Melaleuca's* power to transform the landscape lies in its adaptation to fire. This subtropical evergreen has a thick, spongy inner bark that protects the trunk tissues underneath from fire. In contrast, its flaky and highly combustible outer bark conducts flames upward, igniting its oil-laden leaves. Because lightning strikes are so prevalent in south Florida, opportunities for ignition are rife. After a fire, *Melaleuca* releases millions of seeds to be scattered by wind or water. In this way, *Melaleuca* forms dense stands, either burning or crowding out the already beleaguered native vegetation.[34]

Since the 1960s, *Melaleuca* has experienced a population explosion. The trees now infest 180,000 hectares. In 1983 they were spreading at the rate of 3 hectares per day; by the early 1990s, 20 hectares per day. At these rates the trees, which are four times as thirsty as the native sawgrass, could invade all the wetlands in south Florida in the next half century, turning formerly treeless sawgrass marshes and swamps of cypress and mixed hardwoods to monotypic stands of *Melaleuca*.[35]

Patterning the Landscape/Seascape

Plants that affect the fire regime can obviously alter a landscape quickly and dramatically. Yet plants can also affect the land in more subtle ways, arranging themselves into spatial patterns that change the distribution and flow of water and nutrients and often create a patchwork pattern of productivity. In arid lands in particular, these patterns may stand out starkly as so-called fertile islands or stripes of vegetation across an otherwise barren expanse.

Desert shrubs, for example, often serve as foci for capturing water and nutrients, with each shrub forming an "island of fertility" in a relatively impoverished landscape. In a sandy desert wash in Southern California's Mojave Desert, for instance, investigators found that shrubs and the soil beneath them contained the bulk of the area's nitrogen supply, although shrubs covered only 20 percent of the land area.[36] It turns out that the shrub canopy intercepts rainfall and effectively creates a moister environment at its base, thus providing a favorable micro-climate and soil conditions that encourage greater biological activity. Sediments from the bare ground between shrubs are carried off by water and wind only to be intercepted and deposited around the shrub. Annual grasses and wildflowers cluster near shrubs for shade, moisture, and fertile soil and contribute their own litter to the nutrient stocks stored there. Animals find food and shelter there, too, and add their own droppings and carcasses to enrich the site. Once created, these fertile islands may persist for decades or even centuries.

The formation of such islands in dry regions that historically supported grasslands can be hastened by human-managed activities, such as cattle or sheep grazing, that damage the grasses and encourage shrub encroachment as described in Chapter Four. Because this makes rangeland less economically useful—although not necessarily less biologically productive—excessive encroachment of woody plants is considered a hallmark of land degradation. Once redistribution of resources has occurred, it can be nearly impossible to convert the area to a grassland again, even by resting the land from grazing or ripping up the shrubs and bulldozing the soil. In fact, damaging the shrubs may allow erosion to carry away the topsoil sequestered in these fertile islands, leading to a loss of resources from the ecosystem and causing plant productivity to plummet.[37]

In other arid landscapes, one finds vegetation patterns that resemble stripes, arcs, bands, or ripples. In all these patterns, bands of largely bare soil alternate with parallel bands of vegetation containing some combination of perennial grasses, shrubs, or trees. Near Mecca on the plains of Saudi Arabia, bands of tussock grasses and occasional small trees capture wind and water-borne materials to form ribs of sand. The deserts of Jordan and Syria sport arcs of salt bush. In Mauritania and Niger, *brousse tigree,* or striped bush, is formed by lines of grass and small trees. In Somalia and the Sudan, arcs of grass form in some regions, arcs of *Acacia* trees in others. Micro-ridges topped by sagebrush form ripple patterns in the Utah desert. In western Australia the bands are created by linear *Acacia* groves.

All these landscapes have two features in common. One, they form in climates where rainfall is intermittent and often torrential; and two, they occur on slopes subject to sheetwash runoff. The stripes, arcs, bands, or ripples run at right angles to the sheetwash flow so that each collects the runoff water from the band of bare soil just upslope. The soil conditions in the vegetated bands are better than those in the bare stripes, but only because the plants themselves capture extra topsoil and water and their roots improve the soil structure.

Stripes, like fertile islands, can be easily disrupted—by changes in grazing pressure from sheep or cattle; by plants that colonize the bare areas; or by physical disturbances, such as sheep tracks, that break up the bare-soil areas, allowing water to infiltrate. The result is likely to be severe degradation of the land. In areas of heavier rainfall, overgrazing or other land abuse may drive the expansion of the stripe patterns into

areas now fully vegetated. In the more arid regions, the same disturbances may simply eliminate the vegetated stripes and turn the sheetwash slope to rutted gullies—a type of degradation that has already occurred in parts of the Sudan.[38]

In eastern Australia, too, the water and nutrient-conserving pattern of
the vegetation has essentially ceased to function in many farm pad-
docks. Where linear bands of perennial grasses and *Acacia* (mulga) trees
remain undisturbed, they capture and store almost double the amount
of water they receive as rainfall, and their soils are significantly richer
than those of the sparsely vegetated runoff bands. However, thanks to
intensive grazing and trampling by sheep, rills or gullies now carry water
around or through many of these vegetated bands, threatening their persistence and productivity.[39]

The transitional area where each vegetation band gives way to a barren band is a small-scale version of a larger landscape element ecologists call an ecotone. Ecotones are buffer zones, tension zones between
different ecosystem types or plant communities in a landscape, such as
riparian forests, wetlands, or mangrove swamps. Certain ecotones play
key roles in conserving resources and maintaining the integrity of the regional landscape because they control or intensify the movement of
water, materials, and migrating animals, thus helping to buffer the transition from one ecosystem type to another across the land.

Along inland rivers, for instance, losses or changes in vegetation along
the banks and in the floodplain can dramatically alter the river channel
and the riverine landscape. Quite often, riparian vegetation serves to
capture or filter out sediments and nutrients that might otherwise contaminate the system and be carried out of the region by streams and
rivers. A classic study in a Maryland watershed along Chesapeake Bay
showed that a small band of riparian forest between a cornfield and a
stream reduced by half the nitrate pollutants entering the water. Wetlands, too, play a key role in buffering aquatic systems from the impacts
of terrestrial runoff: they absorb floodwaters, trap sediments, and promote the uptake of nutrients by wetland plants and microbes. Various
wetlands may hold on to anywhere from 14 to 100 percent of the nitrogen they receive, and from 4 to 80 percent of the phosphorus, thus improving water quality downstream.[40]

The ecosystems that buffer the abrupt transition from land to sea are
equally critical. Deltas, marshes, and estuaries sop up nutrients, filter
out sediments, and stabilize salinity levels in areas where fresh water

flows into coastal waters. These natural cleansing and filtering services are particularly important because most of the earth's human population lives in the coastal zone and thus comes into direct contact with the sea. More than 80 percent of the world's fish catch comes from the waters of the continental shelves, coastal margins, and estuaries. Indeed, the highest biological productivity on the earth is concentrated in the coastal plain and offshore zone that runs from two hundred meters above sea level to two hundred meters deep. Yet, 60 percent of the human population lives on the shore side of this zone, and their numbers are projected to double within two to three decades.[41] Can the coastal zone withstand such an onslaught?

Nowhere are coastal services under more pressure than in tropical regions. There the coastal seascape often includes a complex mosaic of wetland mangrove forests, shallow-water seagrass beds, and coral reefs. The lagoons and shallows formed by the interactions of these very different ecosystems shelter some of the world's most productive nursery waters for fish, shrimp, and other sea life.

For their part, coral reefs dissipate the force of ocean waves, tropical storms, and typhoons. Along the land margins, the protective services provided by reefs reduce coastal erosion, and, over eons, encourage the formation of lagoons and protected shorelines where seagrass beds and mangroves can take hold. But the protection is hardly one-sided. The corals that build the reefs require clear water, and sea grasses and mangrove forests quite effectively filter and trap materials pouring off the land that could otherwise cloud coastal waters or smother the corals. In areas where mangrove forests are being cut or where extensive land clearing, storm drainage, or sewage effluents overwhelm the cleansing powers of these ecosystems, the result can be degradation or death of the reefs.

Coral reefs provide a particularly inviting habitat for a rich array of other creatures, and they transform the essential character of the seascape wherever they exist. No other animal can create such monuments. Indeed, the Great Barrier Reef off eastern Australia covers an area some 2,000 kilometers long and 50 to 150 kilometers wide. More stunning is the fact that the architects were not large animals but tiny polyp-like coral animals working in partnership with algae.

Reef-building coral animals get most of their nutrients from a symbiotic partnership with numerous species of single-celled algae called zooxanthellae, which live by the thousands inside the animals' cells. It is the carbon compounds manufactured by these photosynthesizing algae during the light of day that power the yeoman's share of new skeleton-building by the animals. It is the accumulation of these tiny limestone skeletons, created by colonies of millions of interconnected polyps, in conjunction with the cementing activities of coralline algae, that accrete the scaffold of a reef over geologic time.

For reefs to grow, however, the biological construction processes must outrun the forces of erosion that turn reefs to rubble, sand, and silt. An array of grazing animals, from fish to sea urchins, make a living browsing, scraping, or sucking algal growths off the reef, often inadvertently taking up bits of limestone too. Other creatures, such as crown-of-thorns starfish, bristleworms, and certain fish, deliberately bite or scrape off bits of living coral to eat the soft animal tissues inside.

The sea, of course, is a treacherous and storm-tossed place, and natural disturbance has always shaped diversity in coral reef communities. But human activities may increase the frequency of extreme events and also weaken the ability of the reef to rebound—just as we have done in many terrestrial landscapes by speeding up the fire cycle. Fishing practices that use dynamite or poisons to stun fish on the reefs and mining of limestone from intact reefs cause direct destruction. The discharge of sediments and nutrients into coastal waters is indirect but eventually can be just as deadly. Silt and sediments washing over the reef reduce light levels, cut the productivity of the zooxanthellae, and greatly slow the skeleton-building process. High nutrient levels, on the other hand, can not only uncouple the partnership, but also favor growth of smothering layers of filamentous algae. Excessive nutrient inflows might even encourage predator outbreaks on reefs.

Consider the crown-of-thorns seastar. Major outbreaks of this coral-eating animal in the 1960s and 1980s killed large areas of the Great Barrier Reef. Such seastar plagues do occur sporadically even without a human trigger, but there is some evidence that excess nutrient runoff boosts the survival of seastar larvae and sets the stage for a population explosion several years later.

Off the East African coast, where overfishing has depleted creatures that once fed on sea urchin larvae, Kenyan reefs are burdened with one hundred times more urchins than reefs in waters where fishing is restricted. The urchins graze so vigorously that they are eroding away the underlying reef faster than it can be replaced. Yet if a disease were to wipe out this overabundance of urchins, the reef would probably not return to health. Instead, it would likely be covered by algae, because fishermen have depleted grazing fish as well. Such algal overgrowth has already occurred in areas of the Caribbean where algae-eating fish have been depleted and disease has caused urchin populations to collapse.

A 1993 report estimates that 70 percent of the world's coral reefs are either already degraded, likely to be lost within one to two decades, or in danger of disappearing within the next three to four decades. Some of the economic consequences can be anticipated. Reef damage off the Philippines, for instance, has already been blamed for fisheries losses totaling more than $80 million per year and the elimination of 127,000 jobs. In addition, tourism on the Great Barrier Reef generates $1.5 billion each year for Queensland, Australia.[42,43]

Clearly, seascapes and landscapes without corals, elephants, lemmings, bears, or vegetation stripes will change in more than appearance. Ecological functioning will change, and the risk humans face is that these landscapes will not provide the services we require. Early humans, pursuing mammoths across a new frontier, could never have guessed their quarry played a role in shaping the faces of the continents, nor that their own actions could powerfully reshape the new lands they entered. Their descendants can no longer claim such innocence.

CHAPTER EIGHT

Climate and Atmosphere

Overleaf
A major shift in the boundary between taiga and tundra could cause temperature changes throughout the northern hemisphere.

*I*magine a dire scene from the Amazon sometime after 2050. For nearly a century, chain saws and slash fires have pushed steadily through virgin forests and tangled jungle, leveling towering trees hung with lianas and clinging air plants, driving out bats, beetles, toucans, and tapirs. Nature's most extravagant eruption of biological diversity has fallen to human ambition at rates as rapid as one hundred square kilometers a day. Now the neotropical forest is all but gone. The Amazon Basin, covering 40 percent of the South American continent, is one vast tract of degraded pasture.

In the 1990s and even well into the twenty-first century, forest seeds lingering in the soil resprouted quickly in the wake of plows and chain saws, attempting a slow rebirth in the weathered soils of abandoned fields. At some point under the pressure of felling and firing, these attempts faltered. The great forests had reached their point of no-return.

When enough trees had fallen, the Amazon began to turn hotter and drier. Without the dense tree canopy to capture rain and deep roots to draw moisture up from the soil to be vented from leaf pores, the cooling processes of evaporation and transpiration declined. The ground grew hotter. Less vapor was roiled aloft on turbulent convection currents to condense into rain clouds, and a land that once recycled enough moisture to supply half of its own rainfall wilted under dry and cloudless skies. Without the drawing power of masses of moist rising air over the continent, the easterly trade winds that once carried new moisture in from the Atlantic slackened. Rainfall declined dramatically. The dynamic coupling between forest and atmosphere that once sustained a unique regional climate and biotic diversity had vanished with the trees.

Now, lightning fires rage more frequently across the basin, killing any tree seedlings that sprout from the increasingly weathered soils. Attempts to reestablish the moist forests of the past are doomed to wither in the heat of the extended dry season. The Amazon of the late twenty-first century is no longer a suitable place for rain forests.

This scenario played itself out electronically when Jagadish Shukla of the University of Maryland and his colleagues simulated the demise of Amazonian forests on a computer.[1] Yet the interplay of atmospheric and biospheric events chronicled by these researchers would not have raised many eyebrows five hundred years ago.

For it has long been popular wisdom that forests "promote" rainfall. Chop down enough trees and you get less rain. An early biographer of Christopher Columbus claimed in 1571 that the explorer ascribed the regular afternoon showers of the West Indies to "the great forests and trees of those countries." Likewise, Columbus reportedly attributed a noticeable reduction in mist and rain over the Canary, Madeira, and Azores Islands to the clearing of their once-lush forests.[2] Tradition has it that the West Indies later suffered a decrease in rainfall in the seventeenth century after English colonists began clearing most of the rain forests to plant sugar cane. In fact, the belief that the loss of forests was drying the climate prompted legislation in 1791 to create a forest reserve on one of the islands, St. Vincent, "for the purpose of attracting the clouds and rain."[3]

By the late eighteenth century, the topic of just what effect deforestation might have on the weather in North America was a subject that engaged Thomas Jefferson, Noah Webster, and other prominent thinkers. In Europe, French scientists were at the vanguard in investigating the assumed forest–climate connection. The noted French physicist Antoine C. Becquerel even suggested that the climate of Normandy and Brittany might be successfully moderated by eliminating the coastal forests to allow mild ocean winds to sweep inland during the winter. A similar spirit in the United States prompted legislation in 1873 to encourage tree planting on the arid Great Plains as a rain-making venture.[4]

Yet, by the close of the nineteenth century, scientific opinion on forests and climate remained sharply divided, and there were few long-term rainfall or temperature records to help guide the debate. Trees were sometimes assigned contradictory influences by various investigators, and the mechanisms by which they were believed to alter temperature and rainfall remained fuzzy. The debate did little to change the common wisdom, however. American naturalist George Perkins Marsh noted that although no scientific proof seemed to exist, "It has long been a popularly settled belief that vegetation and the condensation and fall of atmospheric moisture are reciprocally necessary to each other."[5] Indeed, as late as 1934, President Franklin Roosevelt was motivated by the notion that trees make rain when he authorized a Shelter Belt Project for afforesting the still-arid Great Plains.[6]

Despite this tradition, the idea fell out of scientific favor, and few modern researchers looked closely at the links between vegetation and climate until it became clear that human activities might be driving large-scale climate changes. Researchers concerned about the profligate release of earth-warming greenhouse gases began a push in the 1970s to develop more realistic computer models of global climate that would help predict the consequences of human alterations of the atmosphere.

To improve the accuracy of their models, one thing researchers needed to know was whether the so-called boundary conditions between earth and air—physical properties determined largely by the plant cover—have a significant impact on climate. These boundary conditions include properties such as the surface roughness created by trees or patchy vegetation; the reflectivity of the surface to incoming sunlight; and evapotranspiration, the combined amounts of water vapor returned to the air by evaporation from wet foliage and transpiration from leaf pores. By the early 1980s, experiments performed with global climate models finally confirmed what folklore had long predicted: regional rainfall, temperature, and wind currents are strongly affected by the type of vegetation that covers the land, particularly its talent for evapotranspiration.[7]

This means that not only does climate dictate whether forests, prairies, or deserts may exist in a region, but a shift in the vegetation can in turn alter the local climate and change the conditions that govern the pattern and type of vegetation on the land. In the tradition of Columbus, many of today's climatologists consider it virtually certain that past human alterations of the land, such as desertification and the clearing of forests, have caused significant changes in weather and climate over some parts of the earth.[8,9]

The fact that climate and vegetation are interrelated has taken on added importance now, not only because of the pace of human land clearing, but also because of the hothouse influence of trace gases, such as carbon dioxide (CO_2) that are now building up in the atmosphere. If the boundaries between such critical biomes as rain forest and savanna or boreal forest and tundra are realigned by global warming, then land-surface boundary conditions could change, leading to further—and possibly larger-scale—climate changes.

Moreover, landscape-level changes in plant cover affect not only the physical conditions at the land surface but also the gaseous composition of the atmosphere, which in turn affects global climate. Through the loss of the carbon-consuming and storing activities of trees, the deforestation of the Amazon would indirectly affect the climate not just of South America but of the globe. Global climate changes, of course, would further realign the earth's vegetation zones, which would drive new changes in climate, and so on in an endless push-pull between atmosphere and biosphere.

Physical Effects of Plants on Climate

Regional climate receives much less attention today than global climate, particularly the impact of atmospheric change on global temperatures. Yet it is the regional scale at which people live, work, farm, and impound water supplies. Whatever change in global averages may occur in coming decades, it is the specific alteration of weather and long-term climate in each region that matters to the human enterprise, as well as to the fate of all other species. It is at this scale—which climatologists call the mesoscale—that the physical effects of plant communities make their mark on climate.

The type of vegetation that covers a landscape influences the continuous exchanges of heat and moisture between the earth's surface and the atmosphere, and it is these exchanges that determine the climatic character of a region—its temperature, rainfall, and wind patterns. On a larger scale, of course, these same transfers of energy and moisture drive global circulation: warm, buoyant air rises, creating areas of low pressure, and cool air sinks, creating areas of high pressure. Air moves horizontally from high- to low-pressure zones, generating surface winds.

Incoming solar energy powers the system, and vegetation plays a leading role in determining the fate of the sunlight that reaches the earth. Some of that sunlight is immediately reflected back to the sky. How much depends on the reflectivity, or albedo, of the earth's surface. Deserts, sparse rangelands, and, of course, blindingly bright expanses of ice and snow reflect much more solar energy back than do dense forests, oceans, and other dark surfaces that have a low albedo. Some of the solar energy that is not reflected away gets absorbed, and dark, unreflective surfaces, such as forest, absorb more than deserts. (This absorbed heat is later reradiated toward space in the form of infrared energy or radiant heat. This outgoing heat is trapped by CO_2, water vapor, and other greenhouse gases, then reradiated downward again to warm the surface, a natural process that, without human intensification, has kept the earth uniquely suitable for life.)

The net solar energy that remains at the surface after these processes of reradiation and reflection represents the energy available to heat the air and soil, fuel green plant photosynthesis, and power evaporation and transpiration (evapotranspiration, for short). Evapotranspiration by plants is a powerful force in the earth's water and climate cycles. As mentioned in Chapter Four, the river water that drains from the continents represents only about one-third of the rain that falls over land. On average, two-thirds of the rainfall is returned to the sky through the activities of green plants and surface evaporation. In some regions and seasons, such as the central and eastern United States in July, enough moisture remains stored in the soil around plant roots so that the moisture returned aloft exceeds the amount rained down.[10]

Evapotranspiration, of course, involves conversion of liquid water to vapor, a task that requires energy and thus draws heat from leaves, soil, and other surfaces, leaving them cooler. (It is this cooling effect that explains why a lushly forested surface covered with transpiring trees is cooler than a desert or pasture, even though the forest has a lower albedo and absorbs more solar energy.)

The vapor released by evapotranspiration may rise from the surface on thermal convection currents or be forced aloft on turbulent updrafts, cooling as it rises. Vapor that ascends a kilometer or so above the surface may cool enough to condense or crystallize into cloud droplets, a transformation that releases the latent heat carried by the vapor into the upper atmosphere, where it drives the global weather engine.

Plants and Regional Climate

Consider in more detail, then, what happened in the computer model when rain forests suddenly vanished from the Amazon Basin. One notable consequence was a marked decline in evapotranspiration. When the modelers looked more closely, they realized that rates of evapotranspiration were closely linked to a number of phenomena; indeed, changes in evapotranspiration seemed to be a key result of many changes that accompanied simulated deforestation.

For one thing, without the trees, the solar energy available to power evapotranspiration dropped sharply. That's because the grassland that replaced the trees had a much greater albedo, reflecting away more than 21 percent of the sunlight compared to 12 percent for the forest. Also, the uniform grass cover intercepted less rain than did the rough and varied surface of the forest canopy, and so less water evaporated from wet foliage before falling to the ground. Without the protection and nourishment of the trees, the tropical soils degraded quickly (see Chapter Five), and their ability to hold water dwindled, allowing more of the rain to run off to the rivers and leaving less available for the plants to transpire. Finally, the shallower roots of the grasses were less able to exploit the limited amount of water that did remain in the soil.[11]

It was no surprise to the modelers that a decline in evapotranspiration would cause an equivalent decline in rainfall over the Amazon. Researchers know that at least half of the rain that falls in the region, and in some areas three-fourths of the rain, is lofted by evapotranspiration to drench the basin again and again.[12]

What was surprising, however, was that rainfall actually declined more than evapotranspiration. That could only mean that the simulated deforestation also reduced the amount of new moisture borne into the basin on humid trade winds from the Atlantic.

At least half of the rain that falls over the Amazon is returned to the air by evapotranspiration from the forest to rain on the basin again.

The key to this finding is that the warm, dry air masses from the deforested basin simply could not rise as vigorously as moist air from the forests. When vapor-laden air rises to the altitude where its moisture condenses out to form clouds, the sudden release of latent heat accelerates the air mass upward, creating a low-pressure column below it that sucks the surrounding air in like a vacuum. With the Andes Mountains forming a wall to the west, the rapidly rising air over the Amazon forest exerts all its pull on the ocean air to the east. Without the kick-start provided by this release of latent heat, the warm air rising languidly from a hotter, drier, deforested Amazon could not exert the same pull on the moist ocean air.[13]

This sort of surprise from increasingly sophisticated computer simulations makes it clear that the relationship between vegetation and climate is not nearly as simple as the trees-promote-rain belief of earlier times. Trees may be important to the total rainfall in most forested regions, but by no means all of them.

In fact, an earlier simulation by Shukla and Yale Mintz of the University of Maryland found that deforestation might actually increase rainfall over India. Just as in the Amazon, stripping the surface of India bare and eliminating the cooling effects of evapotranspiration caused the land to heat up. But geography dictated a different outcome. India is a relatively narrow land surrounded on three sides by ocean. The onshore winds that bring torrential monsoon rains to India are not drawn inland by moist air from evapotranspiration rising over the subcontinent but instead by the phenomenon known as "sea breeze" or the "heat island" effect. Because the land surface heats up much more rapidly in the sun each day than the neighboring seas, rising warm air masses over land create a pressure gradient with the cooler sea air nearby, causing this moist sea air to converge onshore. The model showed, in fact, that if India were deforested, the heat differential between land and sea would increase, strengthening the onshore breezes and drawing in even more moisture.[14] Heavier monsoon rains, however, would likely provide little benefit pounding down on a denuded landscape.

Another example of unexpected—and unexpectedly strong—effects of landscape-level vegetation changes can be seen in models of the boreal forest. Once again, computer models reveal complicated interactions between ocean, atmosphere, land, and vegetation that would have defied prediction.

The boreal forest, or taiga, is a dark, dense band of spruce, fir, pines, and larches that circles the northern reaches of Europe, Asia, and North

America. Compared to the tropical forests, the harsh, remote boreal regions have seen much less human exploitation, although logging operations in the taiga are increasing. To see what significance the loss of these forests would have to regional climate, Gordon Bonan and his colleagues at the National Center for Atmospheric Research in Boulder, Colorado, recently used a global climate model to strip away all taiga vegetation.

Scientists already know that the dark swath of conifers plays a key role in local climate, masking the brightness and reflectivity of the winter snows, absorbing more of the oblique rays of the winter sun, and keeping the surface warmer than it would be otherwise. Thus, when the researchers simulated the elimination of all taiga vegetation, they were not surprised to find that deforestation brought a more bitter chill to the northern winters, dropping the temperature by as much as 12 degrees centigrade during April.

They were surprised, however, by the simulated summer, when the snow melted and the newly bare ground was expected to warm up more than forested ground would under the influence of transpiring trees. Yet the model showed that even in the summer, the barren land remained colder by as much as 5 degrees centigrade. What could account for such a significant climate shift? It turns out that colder winter air temperatures over the land had caused surface temperatures in arctic waters to drop, increasing the extent of sea ice. The ice and frigid waters, in turn, created a holdover cooling effect that inhibited summer warming over the land.

Even when tundra plants rather than bare soil replaced the trees in the model—the equivalent of pushing the tree line between forest and tundra south—both winter and summer air temperatures remained cooler than when the area was forested. Why? Because the short, patchy tundra plants with their sparse foliage could not mask the brightness of the snowpack, and so reflectivity remained high.

The study also revealed that the loss of boreal forests would send a chill into areas well south of today's taiga tree line. In fact, cooler air temperatures would prevail over much of the northern hemisphere—1.6 to 3.2 degrees centigrade cooler year-round at the latitude of Egypt, the southeastern United States, and northern India, and 1 degree centigrade cooler in winter even as far south as the Philippines, the Sudan, and Nicaragua. Indeed, the extent of these hemisphere-wide climate changes, the researchers pointed out, would be "much larger than those due to deforestation in the Amazon."

Just as in the tropical forests, ongoing losses of trees in northern forests could bring the region to a point of no-return, destroying the "coupled dynamical interactions" between vegetation and climate and making regrowth of the boreal forest in its present location unlikely because of the colder summers.[15]

What percentage of either forest could be lost before that threshold was reached? Admittedly, scientists have no answer yet. Both computer simulations described here targeted extreme versions of reality, with all tree cover eliminated to highlight the climate impacts of the vegetation. Thus, neither study gives a clue about the level of deforestation needed to precipitate shifting climate patterns, much less the point where changes would become irreversible. Just as important, researchers have not yet begun to factor into their simulations the indirect effect that forest losses would have on climate through changes in atmospheric gases. Even if deforestation is halted, no one knows whether global climate change brought about by excessive release of greenhouse gases will shrink or expand various biomes or alter the life processes of plants in ways that create unknown feedbacks. These layers of complexity underscore why scientists have turned to complicated computer models to try to analyze and predict future climate.

Unfortunately, modeling of vegetation and its interactions with climate is at an early stage. Most models assume an equilibrium state: If the climate is so many degrees warmer, the model redraws the taiga boundary so many kilometers north. But living communities cannot shift so neatly; a taiga dying back along its southern front as the climate turns hotter and drier may suffer more frequent fires or insect outbreaks.[16] Each of its plant and animal species will migrate at its own pace, according to its own tolerances, with no assurance that the community will ever regroup in its original form.[17] To achieve more realistic predictions, scientists are working to develop "transient," or real-time, models that can track the dynamic interplay of a gradually shifting living community as it adjusts to a gradually changing climate.

In the meantime, by using their computers to look back six thousand years to the middle Holocene, researchers have gotten some notion of the impact a shifting boreal forest might have on temperatures in a climate growing warmer because of the greenhouse effect. In that ancient era, the earth was several degrees warmer than today and the boreal tree line extended north into regions now covered with tundra. Jonathan Foley and his team at the University of Wisconsin found in their simulation that 1.8 degrees centigrade of the high-latitude warming experienced during that period could be attributed to a tilt in the earth's axis. That temperature rise was enough to have pushed the tree line north, and the subsequent expansion of the dark, unreflective taiga canopy then added another 1.6 degrees centigrade to the warming trend. Together these two factors reduced the amount of snow and sea-ice by 40 percent, providing a further boost to the rising temperatures.[18]

Even if the impending doubling of CO_2 in the atmosphere were to cause no increase in greenhouse warming, recent models suggest significant warming over the continents will occur anyway, simply because of the direct impacts of the gas on plant transpiration. Plants grown in control chambers narrow the stomatal openings in their leaves as the concentration of CO_2 in the air increases, allowing less water vapor to diffuse out. Reduced transpiration, of course, means less cooling of the canopy. Even if the world's current tropical forests remain intact, the simulation showed that in a world with twice the CO_2, loss of transpiration alone could significantly warm the region.

In fact, the doubling of CO_2 is much more likely to occur within the next century than complete loss of the tropical forests.[19] Most likely, however, is a doubling of CO_2 along with continued deforestation, and some climate manifestations of the greenhouse effect—a combination with unknown consequences.

Even in the drier areas of the world, clearing of forests and scrublands is already believed responsible for some observed changes in rainfall. A prime example is Australia, where clearing of eucalyptus woodlands and heath scrub in the semiarid southwestern wheatbelt has been followed by a dramatic drop in year-round evapotranspiration and also a decline in rainfall. Between 1950 and 1980, a period marked by extensive land clearing, winter rainfall in this region declined 20 percent. Researchers speculate that the loss of perennial plant cover to annual crops that grow and transpire only during the winter has reduced the supply of water vapor to the atmosphere, increased the albedo, reduced cloud formation over the region, and also allowed moisture that once fell there to be swept further inland.

Researchers investigating this phenomenon are aided by the presence of what Australians call the "bunny fence." The fence is an eight hundred-kilometer vermin-proof barrier erected at the turn of the century in an attempt to keep introduced rabbits from overrunning agricultural lands. Over time the fence has created sharply delineated adjacent regions of grazed and cropped farmlands on one side and remnant native vegetation on the other. The distinctions are clearly visible on satellite pictures. During 1991, dramatic satellite photos taken during the crop-growing season showed convective cumulus clouds roiling up over the native vegetation but absent from adjacent croplands. The explanation, it seems, is that the native vegetation has a lower albedo and so makes more energy available to heat the surface air; once heated, the surface air forms turbulent eddies over the rougher forested surface, introducing updrafts that carry the air masses high enough to condense into clouds. In contrast, the growing crops apparently cannot generate enough convection to lift moist air to heights needed for cloud formation and rainfall. Instead, what vapor the crops transpire is carried away on the surface winds.[20]

Similar phenomena have been proposed for several of the world's great deserts such as the Sahara and the Sinai-Negev, where loss of vegetation to human-managed activities may lead to a decrease in rainfall, which then causes the further loss of plant cover and further drying of the region.[21-23] (Of course, the change in plant cover itself can drive the desertification process, even without a change in local climate patterns, as described in Chapter Four.)

Repeated bouts of suppressed rainfall that rendered the land uninhabitable time and again during the past several millennia have been documented in India, though in this case, the proposed agent is the dust that blows high into the atmosphere from denuded lands. In the Rajasthan Desert, which stretches along the Indus River Valley in western India and Pakistan, intense cropping and grazing have degraded the soils, allowing the winds to churn up massive dust clouds that choke the air. Up to two metric tons of dust per square kilometer hang over the valley today. The dust blocks sunlight, apparently cooling the surface and suppressing rising air masses and cloud formation. Evidence indicates that in previous ages, people and livestock repeatedly abandoned the valley when the rains stopped. In the absence of humans, the grasses eventually returned, then the rains, and after that came the people and their herds to begin the cycle of degradation again.[24]

Plants and the Atmosphere

Only in the past few decades have scientists realized that the composition of the atmosphere plays a central role as a driver of global climate. Because various plant communities differ in their ability to take up or release CO_2 and other trace gases—traits which directly affect the composition of the atmosphere—this has become one of the most heavily scrutinized aspects of ecosystem functioning.

Since the early 1970s, atmospheric scientists have grown increasingly concerned about changes in the levels of certain greenhouse gases generated by human activity. The major greenhouse gases are water vapor, CO_2, methane, nitrous oxide, and manmade molecules known as chlorofluorocarbons. With water vapor excluded, all of the trace gases together make up less than a tenth of 1 percent of the atmosphere. Yet their effect in allowing the passage of incoming sunlight while trapping outgoing radiant heat in the lower atmosphere is profound. Without naturally occurring levels of water vapor and CO_2 in the air, for instance, the earth would be colder by about 33 degrees centigrade, making it virtually uninhabitable for all but the hardiest microbes.

Over the past 250 years, however, as modern industrial societies developed, CO_2 levels have risen from about 270 parts per million to more than 350 parts per million and are continuing to rise at about half of 1 percent per year, thanks largely to increased burning of coal, oil, and other carbon-rich fossil fuels, as well as the cutting and burning of forests. In addition, methane levels from sources such as rice paddies, landfills, and the guts of cattle and sheep, have more than doubled since 1750 and are rising at 1 percent per year.[25]

About one-fourth of the human contribution to rising trace gas levels comes from deforestation, and most of that gas is CO_2 released into the air from soils and vegetation when the trees are cut and burned. Land clearing also releases significant amounts of methane and nitrous oxide, especially when felled trees are burned. Plowing the cleared land for crops over time can release 25 percent of the carbon stored in soil organic matter. Nitrogen fertilizers applied to these fields lead to further emissions of nitrous oxide. Putting cattle on the land or flooding it for rice paddies increases the emission of methane.[26]

An increasing number of scientists believe this ongoing, human-generated rise in greenhouse gas concentrations is already making its mark on the climate. Average air temperatures around the world have risen by half a degree in the past century. While some still argue that such an increase could reflect the natural ups and downs of long-term climate, the United Nations–sponsored Intergovernmental Panel on Climate Change (IPCC) concluded in a 1995 report that "the balance of evidence suggests that there is a discernible human influence on global climate."

The magnitude of eventual greenhouse-driven climate change can only be projected through the use of global climate models, which are still being refined. Nonetheless, numerous simulations using double today's atmospheric CO_2 have led to a consensus by the IPCC that global average temperature could rise somewhere between 1 and 3.5 degrees centigrade by the year 2100.[27] Such climate shifts are certainly not unprecedented in earth history, but the predicted pace of change—which gives ecosystems as well as human societies only decades rather than centuries or millennia to adapt—is unprecedented.

Moreover, models consistently predict large-scale expansions or contractions of various forest types, grasslands, and other of the earth's major vegetation zones—although the models diverge widely on the specifics of the predicted changes.[28] One major set of simulations, for instance, shows the tundra shrinking as boreal forests expand north; the boreal forests in turn lose ground on their southern flank to expanding temperate forests and prairies; the temperate deciduous forests of eastern North America die back to make way for grassland and scrub. The land suitable for tropical forests is projected to expand, although the extent of these forests and the earth's other natural plant communities will depend as much on human plans for the land as on climate.[29]

Carbon Dioxide

The year-to-year rise in global CO_2 levels shows up on a chart not as an upward line but as a steadily rising, methodically oscillating wave. The recurring peaks and valleys in the upward advance of this wave reflect seasonal patterns in photosynthesis as the terrestrial biosphere inhales huge amounts of CO_2 during each spring and summer, then exhales it in a burst of microbial decomposition in fall and winter. (The signal reflects seasons in the Northern Hemisphere because its land area dwarfs that of the Southern Hemisphere.) At the rate the biosphere breathes, an amount equivalent to the total CO_2 in the atmosphere passes through green plant communities on land every seven years.[30]

The carbon that plants pull from the air may be returned to the atmosphere quickly, as microbes or animals that consume fruit, leaves, algae, and other plant material "burn" the carbon compounds to fuel their life processes and breathe CO_2 back to the air as a waste product. Or, the carbon may be sequestered for decades or even centuries in the wood of long-lived trees, in humus-rich soils or peat deposits, or even in layers of dead algae buried in seafloor sediments. The uncertain fate of vast quantities of carbon released to the atmosphere each year adds a major element of unpredictability to global climate change.

Of the 6.3 billion tons of carbon emitted to the atmosphere annually by human activities, the oceans are believed to absorb only about 2 billion tons. Another 3.2 billion tons show up as an addition to the atmosphere. This leaves approximately 1.1 billion tons of carbon "missing" each year, and the suspected consumers that keep this carbon out of the atmosphere are terrestrial ecosystems. If the missing carbon is being sequestered on land, forests are a good place to look for it. Forests already hold most of the land's carbon stores, including 80 percent of the aboveground carbon and 40 percent of the carbon locked up in litter, soils, peat, and roots. Altogether, the global forests contain some 1,146 billion tons of carbon.[31]

Whether the earth's plants really take in more carbon each year than they ultimately release, creating a world that is steadily greener, remains controversial. There is no feasible way to monitor changes in the total standing stock of plant material, above and below ground, across the planet. Yet there is reason to believe that rising CO_2 levels have boosted plant growth at least a bit during this century. Some would like to believe that the vegetation can keep transforming ever greater quantities of carbon into tree rings, roots, leaves, and grain indefinitely, buffering the extremes of atmospheric change and even yielding a lusher bounty of foodstuffs as a consequence.

The fertilizer effect of CO_2 has been known to horticulturalists for at least a century, and many growers of commodities from lettuce to roses routinely pipe the gas into their greenhouses to improve their products. Likewise, hundreds of short-term experiments in greenhouses and growth chambers, mostly with crop plants ranging from soybeans and wheat to oranges, have shown that enriching the air with CO_2 can boost yields by as much as 30 to 40 percent under favorable nutrient, water, and temperature conditions. Yet tests with both crop plants and natural vegetation, including seedlings of forest trees, indicate that the CO_2-induced rise in photosynthesis often levels off with time, or even drops, and some types of plants never experience a growth spurt at all. Moreover, human activities are already injecting far more CO_2 into the atmosphere than the earth's plants are consuming.

One of the clearest messages from studies so far is that some plants will flourish at the expense of others in natural ecosystems where an array of species must compete for access to light, water, nutrients, and other resources.

One of the clearest distinctions in plant responses to CO_2 is based on a species' photosynthetic pathway. Those that stand to gain the most from extra CO_2 are called C_3 plants because they produce a three-carbon molecule during the first stage of photosynthesis. Some 95 percent of all higher plants are C_3, including wheat, rice, beans, potatoes, and virtually all trees. Although the C_3 pathway still predominates today, it evolved in a primordial era of much higher CO_2 levels. Even with today's rising CO_2 levels, these plants still operate inefficiently. Normally, as much as half of the carbon they capture from the air is immediately degraded and lost to an energy-wasting process called photorespiration. When CO_2 levels are ramped up to six hundred parts per million, photorespiration losses drop dramatically.

The group of plants known as C_4 have already made their peace with the relatively low—on a geological time scale—CO_2 levels of the past 65 million years or so. They have evolved a new beginning to the photosynthetic process that turns out a four-carbon molecule and minimizes the energy lost to photorespiration. These plants—a much smaller group but one that nevertheless includes a number of economically important species such as corn, sorghum, millet, pineapples, sugarcane, and many prairie and savanna plants — get little benefit from extra CO_2.

Unless the setting and the soils are right, however, even a C_3 plant may not get much of a boost from extra CO_2. This is illustrated by two of the mere handful of field experiments that have been conducted so far in intact ecosystems. In a nutrient-rich coastal salt marsh in Maryland's Chesapeake Bay, Bert Drake and his team from the nearby Smithsonian Environmental Research Center enclosed C_4 cordgrass and C_3 sedges in open-top chambers and fumigated them with high CO_2 through a number of growing seasons. The sedges received a sustained boost—as much as a doubling of photosynthesis—from this long-running experiment, while the grasses have benefited little. In contrast, a similar test using all C_3 plants but in a colder, nutrient-poor ecosystem yielded different results. When Walter Oechel of San Diego State University injected extra CO_2 into small greenhouses set up over tundra communities of cotton grass and shrubs in Alaska's Brooks Range, he witnessed a growth spurt during the first season, but by the third season, productivity had dropped back to starting levels.

Such isolated experiments, however, predict little about what will happen in real communities characterized by complex interactions among plant species and the animal life they sustain. Two species that each get a boost when grown individually in high CO_2 may get little benefit when grown together—a result that has been found for communities of tree seedlings from both the tropical and temperate forests. Even plants that benefit from increased photosynthesis may use their extra energy in quite different ways: to grow more roots and compete better for soil resources; to grow taller or denser and grab more light; to produce bigger flowers or more seeds; or perhaps to mature, set seed, and die more quickly. The changing rhythms of the plant community may in turn knock it out of sync with the life cycles of pollinators, seed dispersers, or herbivores.

Despite such complicated community dynamics, pot and chamber experiments provide some clues about tomorrow's winners. The discouraging fact is that species diversity is likely to be reduced, and weedy, aggressive C_3 species are in the best position to hold sway, especially if ongoing changes in atmosphere and climate create more frequent disturbances and stress. Some of the winners may even drive new disturbance patterns themselves. For instance, tests show cheatgrass, a notorious invader of rangelands in the western United States (see Chapter Seven), gets a bigger assist from high CO_2 than the native grasses it is displacing. Because cheatgrass naturally intensifies a fire cycle, the prospect of a lusher crop of cheatgrass may mean more and hotter range fires.

Countering Climate Change

Two central questions remain. Will the reconfigured ecosystems of tomorrow consume and store more carbon than they do today? Will their consumption outpace what they release to the atmosphere? No answers are forthcoming, but because increasing CO_2 is likely to be accompanied by temperature changes, the outlook is hardly promising. In a warmer world, some scientists suggest that decomposition and release of carbon stored in soils and litter could outpace the increased uptake by growing plants, at least in the short run, and especially in higher latitudes where warming should be greatest.[32,33]

If the biosphere cannot be counted on to keep up with carbon emissions—and that seems highly likely to most scientists—some people have proposed "geoengineering" as a better solution. One way to do that is by deliberately enhancing greenness, a strategy that on its face seems far less outrageous than other proposed climate manipulations aimed at diverting sunlight from the earth: orbiting giant mirrors, creating artificial clouds, or shading the atmosphere with a blanket of dust fired aloft from naval guns.[34] Even greening the earth would have its pitfalls, whether it was accomplished by smothering lakes in hydrilla and water hyacinth, irrigating the deserts, fertilizing algal blooms in the ocean, or planting massive swaths of new forest.

The idea of greening the oceans has an interesting history. Two oceanographic expeditions near the Galapagos Islands in recent years have confirmed that a lack of the essential nutrient iron prevents algae from taking advantage of the other nutrients abundant in those waters and leaves plant life in the region surprisingly meager. On both expeditions, scientists dripped dilute iron solution into the propeller wash of the research ship across dozens of square kilometers of ocean and were rewarded with a rapid, though short-lived, growth of algae and an increased drawdown of CO_2.

Most scientists profess horror at the possibility of fertilizing millions of square kilometers of the seas with continuous doses of iron—what some wags have labeled the Geritol solution to the CO_2 problem. What ecological effect the iron or the rafts of excess algae might have on ocean chemistry or the ocean food chain is anyone's guess. The notion of deliberately polluting the ocean to counter our unintentional pollution of the atmosphere leaves many deeply troubled—especially when human societies have made little attempt to tackle the underlying problem by reducing carbon emissions at the source.[35,36]

Although the U.S. National Academy of Sciences gave some tacit respectability to the scheme by including it in a 1992 study of possible mitigation approaches to greenhouse warming, the only action that the academy panel was willing to encourage was a reduction in deforestation and a "moderate domestic reforestation program." But just what would it take in the way of new trees to make a dent in the CO_2 problem? The amount of carbon a forest can take up depends, of course, on the species, the soils, the climate, how densely the trees are planted, and how intensively they are managed and fertilized. Two U.S. Forest Service researchers optimistically estimated in 1990 that the nation could sequester more than 56 percent of its own carbon emissions in new wood by planting trees on virtually all "economically marginal and environmentally sensitive pasture and croplands and nonfederal forest lands." The academy panel took a more conservative approach, saying that sopping up 10 percent of U.S. emissions in wood is "a reasonable initial target," although one that would require the diversion of nearly 29 million hectares of often-sensitive lands to intensively managed forests or single-purpose tree farms that provide few other ecological services.[37]

To see what it would take to solve the problem on a larger scale—that is, remove 5 billion tons of carbon from the atmosphere each year, roughly the amount emitted worldwide by fossil fuel burning—Gregg Marland at Oak Ridge National Laboratory in Tennessee did some calculations. He estimated that all the world's 3 billion hectares of closed-canopy forests would have to be spurred to extremely accelerated growth, or 500 million hectares of new trees would have to be grown at the rate of the best tropical forest plantations, or 700 million hectares of new trees with "growth rates comparable to the best of American sycamore plantations on short rotation in Georgia" would have to be planted. To put it in perspective, the last possibility calls for a tree farm nearly the size of Australia. Equally unlikely, all three scenarios assume that deforestation is halted worldwide. As Marland's report points out, there is no law of nature that prevents using trees to solve the CO_2 problem, but "the dimensions of the 'fix' are staggering!"[38]

An additional complication arises; namely, none of these calculations deals with the problematic issue of what is to be done when trees approach maturity and growth slows, or what should happen to the wood. Trees obviously don't grow forever. A mature, natural forest ties up large amounts of carbon, but little new growth takes place except in treefall gaps or burned areas where young, vigorous trees resprout. To lock ever-increasing amounts of carbon away in wood requires harvesting and replanting on much shorter rotations than the complete life cycle of trees. Even then, the wood must be stored long term or turned into durable products, such as houses or furniture. Burning wood simply releases carbon back to the atmosphere. One solution garnering interest is to grow rapidly maturing trees or crops, such as switchgrass, to use as renewable fuels. If such plants were substituted for fossil fuels, the carbon released by burning could be canceled out by regrowing plants, causing no net increase in atmospheric CO_2. By one estimate, short-rotation tree crops have the potential to reduce the world's fossil fuel emissions by as much as 20 percent.[39]

Even if CO_2 emissions are stabilized, however, other greenhouse gases may continue to rise. Two of these trace gases, methane and nitrous oxide, are hardly insignificant: they are thought to account for 20 percent of the predicted greenhouse warming.[40] Recent findings about the indirect effects of rising CO_2 on emissions of these other gases should give the backers of green-up solutions to global change some second thoughts. Some of the extra plant productivity fueled by increasing CO_2, especially in nutrient-rich wetlands, may come back to haunt us in the form of extra methane, a much more potent greenhouse gas. Any schemes that call for spreading more nitrogen fertilizers across fields and forests to encourage plants to draw down extra CO_2 may in turn yield up more methane and more nitrous oxide.

Methane

Methane is second only to CO_2 as the most important human-generated greenhouse gas. Although it exists at atmospheric concentrations 175 times lower than CO_2, a molecule of methane is twenty times as effective in absorbing and trapping radiant heat. The sources and sinks in the methane budget are not well quantified, but human activities such as intensive livestock grazing in temperate and tropical grasslands and flooding of large areas for rice cultivation are believed responsible for doubling atmospheric methane over the past 250 years. Since the late 1970s, methane concentrations have been rising at about 1 percent a year, twice the percentage rise of CO_2.

Soil microbes serve as both the producers and consumers of methane, although the amount that actually reaches the atmosphere depends on a complex web of interactions between microbes and plants. Production of the gas is controlled by approximately one hundred species of bacteria called methanogens that ferment organic material such as plant litter, degrading it into methane gas. Methanogenic bacteria do their fermentation work in anaerobic, that is, oxygenless, environments such as the mud of wetlands, rice paddies, landfills, and even the intestinal tracts of termites and cattle. Biomass burning, coal mining, and natural gas drilling (methane is the major component in natural gas) also release methane to the atmosphere.

Another group of bacteria called methanotrophs, which number only about twenty-five species, intercept and consume the gas in nearby oxygenated soils, using it to fuel their own life processes and releasing waste CO_2. This way they release a less potent gas to the atmosphere. These methane-consuming bacteria are common in most soils, from forests to prairie grasslands, and in some drier regions they consume more methane than these soils produce, turning these areas into net sinks for methane.

Plants do not make or use methane, but they have two influences on how much reaches the atmosphere. First, plant material provides the carbon that is broken down and reformulated by methanogens to become methane. And second, wetland plants serve as conduits for methane from the deep soil to the atmosphere. It turns out that these plants have a system of internal air spaces that supplies oxygen to roots buried in the anoxic mud and also allows transport of methane upward into the air.[41]

Until recently, more effort has been put into identifying the sources of methane emissions than into studying methane consumers, yet the historical buildup of methane may reflect a decline in consumption as well as increased emissions.

Nearly one-fourth of the human-caused emissions of methane come from flooded rice fields, although changes in management can sharply reduce this source of emissions. One such strategy calls for draining rice fields. Doing so once per season replenishes oxygen supplies to the paddy soils, thus eliminating the anaerobic conditions that are ideal for methane producers and cutting methane emissions almost in half. Draining the fields every three to four weeks throughout the growing season can cut emissions by 88 percent without affecting rice yields. (Selecting the right strain of rice can also make a difference. Tests in India and China have shown that emissions can more than double from one strain of rice to another growing under the same conditions.)[42,43]

Drier soils seem to be a consistent favorite with methane consumers around the world. This fact has important implications for climate change projections. For example, because today's moist tundra provides a significant amount of methane to the atmosphere, many simulations of climate warming have predicted these emissions will increase as the peat and permafrost begin to thaw. None of these models take methane consumers into account, however. Yet field experiments on a wet tundra meadow in the Aleutian Islands suggest that a warmer and drier climate will cause the water table in many parts of the tundra to drop, giving methane consumers a real boost. Some researchers now speculate, in fact, that the tundra of tomorrow could turn into a net sink rather than a source for methane emissions.[44]

Still, such a dampening effect may be counterbalanced by the expansion of modern agricultural landscapes, which may put out more methane than they consume. Certainly the human-caused rain of nitrogen that is affecting broad areas of the planet seems to be indirectly driving up levels of atmospheric methane too.

For example, when nitrogen fertilizer was applied to temperate forest soils in New England—bringing the nitrogen up to levels common in heavily polluted forest areas of western and central Europe—methane consumption by microbes fell by one-third. That's twice the decline caused by wetter soils.[45] A similar effect was seen on the Colorado prairie, where plowing and fertilizing the natural grassland soils decreased the uptake of methane.[46]

These findings in forest and prairie are echoed by those from one of the longest-running agricultural fertilization experiments in the world, the wheat fields of Rothamsted Experimental Station in the United Kingdom. After 140 years of continuous inorganic nitrogen applications, methane consumption by soil microbes has been greatly reduced. In short, the higher the doses of fertilizer a field has received, the greater its methane emissions to the atmosphere. (A decline in methane consumption has not occurred in fields enriched with manure rather than inorganic nitrogen, probably because the carbon-rich manure allows the soil to support larger populations of microbes, including methanotrophs.)[47]

A still bigger determinant of methane emissions may be plant productivity, because it fuels the work of methane producers. Indeed, Gary Whiting of NASA's Langley Research Center in Virginia and J. P. Chanton of Florida State University consider productivity a "master variable." To find out what is responsible for the large differences in methane emissions from various types of wetlands, the two researchers traveled from fens and bogs in Canada and Minnesota to cattail swamps in Virginia, sawgrass marshes in the Florida Everglades, and rice paddies in Louisiana, measuring both methane emissions and the uptake and release of CO_2 (a measure of productivity). In all those settings, about 3 percent of each day's net plant production was decomposed by methanogenic bacteria and emitted as methane. Thus, the more abundantly a marsh produces, the more gas gets released.

The finding has obvious implications for climate change, since rising CO_2 levels should fertilize at least some degree of extra plant growth in nutrient-rich ecosystems like wetlands. A greener bog may take up more CO_2, but it may also produce more methane and create a positive feedback that further enhances the greenhouse effect.[48]

That prognosis recently got a boost from the long-term CO_2 enrichment study mentioned earlier in a coastal marsh on Chesapeake Bay. It turns out that C_3 sedges growing with extra vigor in double the present amount of CO_2 were also emitting 80 percent more methane than unfumigated sedges growing nearby.[49]

Nitrous Oxide

Nitrous oxide is a very long-lived greenhouse gas whose production has nearly doubled in the past century. Some of this rising tide of nitrous oxide is generated by automobile exhausts and the burning of other nitrogen-loaded fossil fuels such as coal, but a major portion of it comes from the microbial breakdown of ammonium or nitrates in the soil.

Production of nitrous oxide is particularly steep in agricultural areas where high rates of nitrogen fertilizer are applied to the soils, and in ecosystems like temperate forests where, ironically, acid rain and dry deposition (some from nitrous oxide generated elsewhere) deposit high loads of nitrogen pollutants. The same studies in the Colorado prairies that saw methane consumption fall as the land was plowed and fertilized, for instance, also reported that nitrous oxide production doubled or tripled under these intensive farming practices.[50]

Soils in the tropical rain forests may account for a large proportion of atmospheric nitrous oxide. Amazonian forest soils have been found to release nitrous oxide at a rate twenty times the average for other soils worldwide.[51] Yet clearing and burning tropical forests for pastures may actually lead to emissions five to eight times as high, at least for the first decade or so after deforestation. Studies in the Atlantic lowlands of Costa Rica indicate that these emission levels peak during the first decade, and after two to three decades a mature pasture may produce less nitrous oxide than the forest. Rather than a cause for comfort, however, those reduced emissions seem to be a side effect of the nutrient depletion and soil degradation that accompanies deforestation. The debris and soil organic matter inherited from the forest are broken down rapidly at first, and since the grasses cannot use all of it, excess nitrogen is lost to the atmosphere. With time, the tropical soils become so nitrogen-poor that nitrous oxide emissions drop below those of the original forest.[52]

Dimethylsulfide, Clouds, and Gaia

A compound generated by living organisms that may help to counter the effects of greenhouse gases is dimethylsulfide (DMS), which apparently plays a major role in the formation of earth-cooling clouds over the oceans. Many types of marine algae produce the parent compound to DMS. Only a few groups of algae, however, produce and emit it at high rates. Emissions can be boosted dramatically by tiny zooplankton grazers that feed on the DMS-producing algae and expel feces containing high concentrations of this sulfur compound.[53] Aloft, DMS reacts to form sulfate aerosol particles, which then serve as "seeds" for cloud formation. As moist air rises and cools, its vapor must condense around so-called cloud condensation nuclei—often dust, salt, or other impurities in the air. Over the oceans, DMS turns out to be the major source of such nuclei, and thus algae indirectly influence the type and abundance of clouds over the seas.[54]

In general, the more of these nuclei available, the brighter and more reflective a cloud will be, bouncing more sunlight back to space and helping to cool the earth's surface. Thus, any ups or downs in plankton populations that affect DMS emissions could alter cloud cover—in particular, the thick, low-lying banks of marine stratocumulus clouds that are known to have a net cooling effect on the earth. (Some terrestrial algae also emit DMS, but the resulting particles are dwarfed in number by the enormous volume of condensation nuclei supplied by human-generated pollutants such as sulfur dioxide.)

The very fact that oceanic algae might have a hand in regulating the temperature of the earth brought these DMS-producers to prominence in the late 1980s in the debate over the controversial Gaia hypothesis. This hypothesis, first put forward by James Lovelock in the early 1970s, suggests that living things have the power to regulate environmental conditions such as surface temperatures and the mix of gases in the atmosphere in ways that keep the earth livable. In other words, the earth itself functions like an organism, self-regulating and self-correcting, although adjustments might require millions of years.[55]

Gaia supporters reasoned that some sort of negative feedback loop would drive DMS-producers to bloom and emit more cloud-forming particles to help cool the surface when the climate warms to unacceptable levels. So far, no evidence of such a feedback exists. Scientists don't even know whether warmer seas would spur growth of DMS-producing algae, or whether cooling seas would depress algal growth and lead to cloudless skies.[56] Further, proponents of Gaia have not been able to devise models of any feedback loops strong enough to prevent a warming climate from growing ever hotter.[57]

Indirect human influences, such as ozone depletion and the increase in ultraviolet radiation reaching the sea, may alter these algal populations in unpredictable ways. The link between numbers of algae, the amount of DMS they produce, and changes in cloud abundance remains uncertain, making it impossible to predict what level of change in plankton populations might significantly affect cloud formation or climate.[58]

Obviously, the notion that the earth is self-regulating and equipped with a thermostat that will eventually pull the climate and atmosphere back into some happy configuration is an attractive one, though no evidence for it exists. Lacking such evidence, it seems especially ludicrous for human beings to stand passively in the face of global changes we have unwittingly set in motion and hope that something will spur algae, sedges, or spruces to effect our rescue. The biological feedback mechanism that has the best chance of keeping the planet in an amenable state is our own species' seldom-exercised power of self-control.

CHAPTER NINE

Do We Still Need Nature?

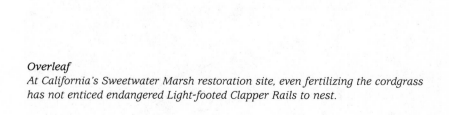

Overleaf
At California's Sweetwater Marsh restoration site, even fertilizing the cordgrass has not enticed endangered Light-footed Clapper Rails to nest.

*I*n 1991, after seven years of planning by engineers, architects, and ecologists, eight men and women stepped into a self-contained, synthetic world in the Arizona desert and closed the airlocks. Joining them inside their monumental glass and steel enclosure called Biosphere 2 were three thousand other species, a carefully chosen subset of the earth's biological diversity organized into facsimiles of five major biomes. At a cost of $150 million, Biosphere 2 was the most elaborate artificial life-support system ever designed, a far cry from the thirty-liter tank of green algae that had kept a Russian life-support pioneer alive for a single day three decades earlier by supplying him oxygen and taking up his exhaled carbon dioxide (CO_2). The goals of Biosphere 2 were as grand as its price tag: It was intended not only to nurture and sustain its human crew for two years, but also to "maintain resilient, persistent, complex, and evolving ecosystems."

Despite some successes during its two-year run, Biosphere 2 dramatically illustrated just how difficult it is to assemble or reassemble fully functional ecosystems, even when nature supplies a diverse choice of components.

The design called for a maximal array of species, and so ecologists involved in the project opted to recreate a vast sweep of the earth's midsection from equatorial rain forest to subtropical fog desert. They compressed it all into 168 meters to fit inside the one-and-a-quarter-hectare enclosure. The tropical rain forest was stocked with more than three hundred Amazonian plant species, including eighteen-meter-tall *Leuceana* trees that soon had to be pruned away from the dome. A coral reef was flown in from the Caribbean to grace the miniature ocean, which lapped a mangrove swamp on the border of a cattail marsh. A grassy savanna with African *Acacia* trees and seasonally flooded pools called billabongs was created alongside a fog desert patterned after the Baja California coast. Each biome had its own team, and each chose its own strategy. The wetland group extracted chunks of real marsh from the

Everglades and trucked them intact to Arizona, while the desert team assembled a mix-and-match array of boojum trees and cordon cacti from Baja interspersed with exotic succulents from Israel, Namibia, and Chile. Together the five "wilderness" biomes shared the dome with human habitat areas and a working farm containing a vegetable garden, orchard, rice paddies, goats, and chickens.

Although the ecologists designed their plant assemblages with great care, they gave little thought to gas exchange and how the sealed atmosphere might interact with the biological activity below. This one oversight proved to be disastrous.

Almost immediately after the dome was sealed, CO_2 rose to dangerous levels, three to seven times higher than outside; also, oxygen concentrations dropped from 21 percent to less than 15 percent. In addition, the relatively tiny volume of air was overwhelmed by large daily swings created by Biosphere 2's metabolizing organisms. Consequently, CO_2 levels fluctuated drastically, with photosynthesizing plants absorbing it during the day and the entire biota exhaling it at night.

The unexpectedly large exhalation of CO_2 was mostly the work of microbes in the agricultural soils. Wanting to maximize fertility, the designers had filled the garden plot with rich organic soils characteristic of temperate grasslands. But under the moist tropical conditions of Biosphere 2, microbes in the soil went into overdrive, decomposing organic matter at a rapid pace, using up oxygen as they worked, and releasing more CO_2 than even the dense array of plants in the dome could use.

As CO_2 levels soared, the ocean pH became acidic and the corals began to dissolve, forcing the crew to add bicarbonate to buffer the waters. The fresh water supply, constantly recycled through the vegetation and soils, grew increasingly saltier. With the rain forest already growing at full tilt—or as much as it could in the low-light months of a temperate winter—the crew pruned the plants to encourage more carbon-consuming growth. They ceased most composting and began to store massive quantities of plant clippings in Biosphere 2's basement to prevent decomposition and the release of more carbon. They forced the savanna to keep growing, depriving it of a dry season. They also doubled, then tripled rainfall on the desert to spur plant growth. The extra rainfall, however, enabled savanna grasses to invade the desert, causing a third of the desert species to die out.

Despite all these efforts, conditions in Biosphere 2 continued to deteriorate. Most of the insects, including bees and other pollinators, perished, leaving the crew to pollinate their corn by hand. Yet pestiferous insects thrived, reducing garden production and leaving the crew's diet lean; each lost an average of twenty-five pounds during the two years inside. The crew members also found themselves culling predatory lobsters and parrotfish from their ocean, picking seaweed off their reef, pulling weeds from their wild biomes, and enduring outbreaks of cockroaches and crazy ants as they labored to sustain the designer ecosystems that were supposed to sustain them. Engineers had to install CO_2 scrubbers and begin infusing oxygen into the dome just to keep the Biosphereans healthy.[1-3]

Today, under new management and devoted to research and teaching, the Biosphere 2 facility no longer sustains humans in a synthetic environment. Its biomes are still growing, its species sorting themselves into communities that may yet reveal some of the intricacies of how ecological systems function in the real world. Such knowledge may prove useful in two ways: to aid in the restoration of degraded natural systems and to help place a realistic value on sustaining the earth's remaining biological diversity. Until human societies can understand in both ecological and economic terms just what is lost when we abandon or degrade the free subsidies nature provides, we are likely to continue bartering away too cheaply the essential elements of our life support systems.

Ecological Restoration

The notion that degraded lands can be restored to a fully functional and self-sustaining condition has gained popularity in some circles, providing hope that much human-caused ecological damage can be undone and natural landscapes recreated. This hope has spawned a fledgling field called restoration ecology, a branch of ecological engineering.[4] So far, restoration ecology has yet to produce a clear-cut success, and most projects are still works-in-progress.

The longest-running restoration efforts have been conducted on the North American prairies, where only a fraction of 1 percent of the original prairie biome remains after more than a century of fire suppression, farming, grazing, and invasions of exotic species. Hundreds of hectares of prairie vegetation on restored sites now look much as they must have when the first European pioneers reached the Midwest. The most ambitious project underway can be found on the grounds of the Fermi National Accelerator Laboratory in suburban Chicago. Although restoration work has concentrated on regaining the proper mix of native plants on the 240-hectare site, some ecological functions seem to be improving too. For instance, the structure and water-holding capacity of the soils have improved dramatically compared to nearby land planted in pasture grasses.[5]

Yet only recently have ecologists attempted to re-create the full complexities and functioning of natural ecosystems. Such restoration projects may take the form of accelerating the natural recolonization process. Ecologist Dan Janzen of the University of Pennsylvania, for instance, is attempting to encourage the regeneration of tropical dry forests—a nearly vanished ecosystem in Central America—on lands in Costa Rica's Guanacaste Province that were long ago cleared for agriculture. Other ecologists are trying to speed up regeneration in riverine areas damaged by hot-water discharges from power plants.

In the United States, much attention is being given to efforts at artificial wetland construction. Since the 1980s, extensive wetland restorations have been attempted throughout the nation thanks to a policy of "no net loss" that allows developers to destroy wetlands at one site by agreeing to improve degraded wetlands or create equivalent new wetlands elsewhere. Thousands of ill-conceived and largely unsupervised mitigation projects have been carried out, often with little monitoring or followup. Meanwhile, thriving natural wetlands have been bulldozed, drained, and filled to make way for construction.[6]

Creating something that looks like a wetland, that contains the right water levels and the right plants for the region, seems easy enough, but putting all the pieces in place has proven to be surprisingly difficult. When workers bulldoze a pond in the uplands and plant appropriate reeds or cordgrass, the replica will often perform some key functions: buffering floods, allowing sediments and nutrients to settle out, and even providing a rest stop for easy-to-please waterfowl. Indeed, over the past decade, ecological engineers in the United States and Europe have had some success designing wastewater-treatment systems that mimic this function of natural wetlands. Those projects let intense, solar-fueled activities of soil microbes and marsh plants convert the organic and nutrient content of sewage into biomass rather than rely on mechanical treatment systems to cleanse the water. Creating new wetlands that work well enough to support a full range of wetland plant and animal species is a different matter altogether.

One exceptionally well-monitored project in southern California, where less than 15 percent of the historic expanse of tidal marsh remains, illustrates just how far from complex natural functioning an otherwise authentic-looking wetland can be. Restoration at the site, known as the Sweetwater Marsh National Wildlife Refuge, represents a federally mandated effort to provide suitable habitat for two endangered birds and also to reestablish populations of an endangered plant.

Joy Zedler of San Diego State University began tracking progress and conducting research in the new marsh in 1989, five years after it was created from former marsh and mud flats long buried under urban trash and dredge spoils. Her team successfully seeded the endangered plant, the salt marsh bird's-beak, into the site. Small native fish that provide food for one of the endangered birds, the California Least Tern, have returned to the reconstructed channels. The terns, however, rarely fish there, for reasons still unknown.[7] The other endangered bird, the Light-footed Clapper Rail, has proven even harder to please.

Clapper rails require tall cordgrass to camouflage their floating nests from predators and allow the nests to rise on the tide without being carried away. The new marsh grows dense cordgrass, but it cannot seem to grow it tall enough to attract the birds.[8] The problem, it seems, is that the coarse, sandy soil from the old dredge spoils contains little clay or organic matter and so can't store the nutrients needed to grow tall grass. Kathy Boyer of Zedler's team has tried fertilizing the marsh, but only by fertilizing it every two weeks throughout the growing season did she achieve the grass height preferred by the rails.[9] Even then, the rails shunned the new marsh, as they have for more than a decade.

The restoration project is still underway, and the final answers are not in yet. Indeed, the outcome of projects like these may not be clear for decades or longer, well after the usual five-year period for monitoring such mitigation projects has passed. Creating suitable habitat and recovering lost functioning has proven a tough assignment, and not just in Sweetwater Marsh. When Zedler examined coastal mitigation projects throughout southern California, she found no instance where ecosystem functioning had been successfully duplicated or where endangered species had been "rescued from the threat of extinction."[10] A 1992 report by the U.S. National Academy of Sciences concluded much the same for all types of wetland restorations throughout the nation: There is no evidence that these restored wetlands can re-create fully functional ecosystems, and they cannot be relied upon to maintain biological diversity.[11]

Certainly, restoration efforts to return previously damaged lands to ecological service should be encouraged. The National Academy report itself called for restoration of 4 million hectares of already degraded wetlands in the United States. There are millions more hectares of rangelands, forests, and marsh on every continent that might be returned to health and productivity if complex environments could be rebuilt as skillfully as they are being dismantled. As for the earth's dwindling stock of natural and seminatural landscapes, it would be far less costly to preserve robust natural systems rather than count on being able to piece them back together after they've been torn apart. Certainly, the current track record of restoration offers nothing on which to stake the fate of the earth's functioning ecosystems.

An Engineering Fix

Most industrial societies tend to disregard and devalue ecosystem processes, opting instead for a technological fix whenever environmental services falter. Lost services are replaced not with natural mimics but with engineering solutions: dams, reservoirs, waste treatment plants, air scrubbers, air conditioners, synthetic pesticides and fertilizers, and water filtration systems. Replacing natural systems that work for free with engineered systems powered by fossil fuels is enormously expensive, often problem-plagued, and commonly impractical, especially in developing countries. Furthermore, these systems seldom attempt to replace the full array of services supplied by the natural system.

Herbert Bormann of Yale University pointed out two decades ago a litany of services that must be replaced when forests are cut:

> We must find replacements for wood products, build erosion control works, enlarge reservoirs, upgrade air pollution control technology, install flood control works, improve water purification plants, increase air conditioning, and provide new recreational facilities. These substitutes represent an enormous tax burden, a drain on the world's supply of natural resources, and increased stress on the natural system that remains. Clearly, the diminution of solar-powered natural systems and the expansion of fossil-powered human systems are currently locked in a positive feedback cycle. Increased consumption of fossil energy

means increased stress on natural systems, which in turn means still more consumption of fossil energy to replace lost natural functions if the quality of life is to be maintained.[12]

Ecologists Eugene P. Odum of the University of Georgia and Howard T. Odum of the University of Florida were among the first to try to quantify the economic value of self-maintaining ecological systems. By the early 1970s, they anticipated that "neither aesthetic values nor ethnic traditions" would provide an adequate incentive for preserving natural environments because of the development pressures being generated by rapid technological and population growth. Yet these same pressures would also create an even greater need for the free sun-powered services of nature in order to help maintain the quality of life in growing cities and other "power-hungry" developed areas. The Odums calculated that even the "per capita cost of treating human wastes, which are only one small part of the pollution disorder generated by cities, would be more than doubled if there were no natural environment available and able to carry out the work of tertiary treatment of these wastes."

The Odums devised an admittedly oversimplified model for determining how much natural land it takes to support a city. Until this kind of analysis can be refined, they suggested, "it would be prudent for planners everywhere to strive to preserve 50 percent of the total environment as natural environment."[13] In their model, the boxes representing natural lands and ecosystems contained no specifics about forests, wetlands, or grasslands, much less lists of species supplying services to the "developed" boxes. If the earth were diagramed into such a model today, it would be obvious that the size of the boxes on the developed side has ballooned over the past twenty-five years while the size of those on the natural side has shrunk rapidly. Yet ecologists still have few names and numbers to put in those shrinking boxes. Furthermore, economists are just beginning serious efforts to quantify and assign monetary values to the input and output arrows running between the natural and the developed worlds.

Putting a Price Tag on Ecological Services

Conventional economists have attempted to include biological diversity in their calculations by assigning monetary figures to the various types of values species hold for the human enterprise: commodity values, based on their worth in the marketplace or in generating tourist dollars; amenity values, based on their aesthetic or nonmaterial worth to humans; and moral values, either based on their moral worth to humans or, some would say, their own intrinsic moral worth as living creatures. Economists have even drawn into their analytic framework species whose worth as commodities or even amenities is currently low or unknown. They do this by assigning species option values, a sort of surrogate worth based on the possibility that some use or real value will be conferred on them in the future.

None of these categories, however, gets directly at what it is worth to have species work together within ecosystems to generate the life-support services that make the earth habitable. Clearly the degradation of ecosystem functioning has an impact on human welfare and economic activities that goes beyond the value of the timber or other commodities gained or lost. "The value of biological diversity," as Bryan Norton of the Georgia Institute of Technology wrote, "is more than the sum of its parts."[14]

So far, economists, resource managers, and policy analysts have made only tentative efforts to factor into land-use decisions the value of ecological services that will be impaired when a forest is logged or a marsh drained. Thomas Lovejoy of the Smithsonian Institution illustrated the difficulties of trying to integrate the ecological with the economic world view:

"How should the American oyster population of the Chesapeake Bay be valued? Is its value what it brings to market as seafood annually? Or is the value that the current population filters a volume of water equal to the entire bay once a year, and its value before degradation of the bay that it filtered that same enormous volume once a week? Our economies are riddled with such beneficial subsidies from nature, for which there is no current accounting. Similarly, our economies are riddled with subsidies and incentives that lead to environmental degradation."[15]

Assigning value to the full range of ecological services provided by the natural world may prove impossible. Still, the known economic value of certain services to agriculture such as nitrogen-fixation, pest control, and pollination is staggering. Nitrogen-fixing microbes in the soil supply $7 billion worth of nitrogen to U.S. agriculture each year. Worldwide, microbes in agricultural soils annually fix an estimated 90 million tons of nitrogen, worth nearly $50 billion. Insect pollinators provide an absolutely essential service to many crops: David Pimentel of Cornell University and his colleagues report that more than forty U.S. crops worth $30 billion are completely dependent on insect pollinators for their productivity.[16] Yet often, a pollinator's full worth to a specific crop or region is apparent only when it is missing or added.

A case in point involves a weevil that pollinates African oil palms. The oil palm was introduced into Malaysia from the forests of Cameroon in West Africa in 1917; the weevil was not. Palm growers in Malaysia had to rely instead on expensive and labor-intensive hand pollination. Then in 1980, the weevil was captured and imported to Malaysia. The weevil soon boosted fruit yield in the palms by 40 to 60 percent, and also generated savings in labor costs of $140 million a year.[17,18]

People throughout the world spend about $20 billion a year for pesticides, yet the pest control services provided by predators and parasites that thrive in natural ecosystems are worth an estimated five to ten times that amount. Without this subsidy from the environment, Pimentel points out, both pesticide costs and crop losses would soar.

One of the few specific calculations available focused on the pest control work of the *Anolis* lizards of the Antilles islands in the Caribbean. By preying on insects that attack sugar cane, bananas, and cocoa crops, the islands' primary export commodities, the lizards reduce the demand for pesticides. Estimates showed that complete elimination of the lizards from the islands could cause food production losses as high as $455 million. Even a 1 percent drop in the lizard population might cost $670,000 in reduced yields.[19]

Another service provided by natural ecosystems is that of preserving novel genes of incalculable value. For example, seeds taken from berry-sized, green-and-white striped fruits of a wild tomato species discovered high in the Peruvian Andes in 1962 have been hybridized with commercial tomato varieties to produce new strains with higher sugar content and better flavor, a development that has increased the value of California's tomato crop by $8 million a year.

The value of that weedy tomato will probably be dwarfed by the billions of dollars in benefits expected from the genes of a wild species of corn found in the Mexican state of Jalisco in 1977. This wild type is the only known perennial species of corn; it is also exceedingly resistant to the major diseases that plague domesticated corn. Plant breeders throughout the world are still exploring ways to incorporate the genes of this find into commercial varieties of corn, which is the world's third most valuable crop.[20]

It stands to reason that wild ecosystems offer immense "option value" to society. Yet putting a realistic dollar value on the services they provide as living repositories for potential crops, medicines, and other products is a daunting task.

One problem with valuing and protecting services such as genetic diversity, flood control, waste clean-up, and even natural pest control is that the benefits typically extend well beyond the boundaries of defined ecosystems. In other words, the ecosystem services threatened by a particular land use decision may provide regional benefits, while the land use change gives direct value to the landowner. Society has not yet dealt with the question of requiring landowners and other resource users to internalize the cost of degraded or lost ecosystem services in the way that industries, municipalities, and even automobile owners in recent decades have finally had to internalize certain pollution costs by paying for scrubbers, filters, and converters that limit their emissions to the air and water.

The problem is illustrated by a recent analysis of how salmon production in the coastal estuaries of Sweden would fare if held to the same sewage treatment standards as cities. These farming operations rear salmon in floating cages and rely on large tidal flows to flush away fish wastes and replenish oxygen levels in the cages. Besides generating huge amounts of waste, intensive management of the fish includes dousing them with antibiotics, chemicals, and supplemental feed. The inputs to estuarine and coastal waters can be so high that the farming operations overwhelm and degrade the cleansing function of the estuary, leading to eutrophication. Yet the aquaculture industry is not usually required to treat its wastes the way cities and industries must treat their sewage effluents.

Carl Folke of Sweden's Beijer International Institute of Ecological Economics calculated that if the "external" costs of increasing eutrophication in coastal waters were included in the expenses of salmon farming, the cost of production would rise by about 75 cents per kilogram of fish, bringing the total cost to about $5—higher than the market price of farmed salmon. In short, if fish farmers were required to pay the full cost of water purification, the industry probably would not be profitable or sustainable. Furthermore, this analysis only considers water quality.

Salmon-farming operations in estuaries may generate enough wastes to over-whelm the cleansing function of these wetlands.

Caging fish at high densities can lead to disease and parasite problems that may spread to surrounding plants and animals, including wild stocks of commercially important shrimp or fish. Escaped fish may hybridize with wild stocks, perhaps altering the genetic makeup of these populations in ways that reduce their long-term survival. Putting a monetary value on these impacts and including them in the cost of production would be problematic, but would certainly make salmon farming less profitable.[21]

Recently, ecologists in South Africa tackled a complex question: Can the ongoing expense of keeping alien plants from overtaking the native vegetation of the Cape region be justified economically in a nation faced with an overwhelming backlog of human needs?

The basic problem is that proteas and other shrubs of the fynbos, or scrub, community are being crowded out in many parts of the Cape by dense groves of invading trees, especially Mediterranean pines, Australian *Acacias,* and others imported by European settlers to afforest the treeless scrub. The tree invasion has a recognized economic cost: trees obviously use more water than shrubs, and so reduce water yield from the mountainous catchments. But keeping invading trees under control on the slopes has an economic cost, too. It requires a continuing regimen of hand-pulling and chopping, combined with well-timed burns, to keep the trees from taking over. Does the amount of water saved balance the expense of keeping the fynbos system intact and functioning?

Brian van Wilgen of the CSIR Division of Forest Science and Technology, Richard Cowling of the University of Cape Town, and Chris Burgers of Cape Nature Conservation analyzed the tradeoffs and concluded that protecting the fynbos system makes good economic sense. According to their models, the intact fynbos delivers more water, more cheaply, even factoring in the cost of fighting weedy species.

The three began their analysis with the results of a computer simulation, which indicated that over one hundred years, alien plants left uncontrolled would cover 80 percent of a watershed and result in "average losses of more than 30 percent of the water supply to the city of Cape Town." The group then compared the costs of developing new water supply facilities in a tree-filled watershed versus a fynbos watershed where invading trees are kept in control. Obviously, the cost of building and operating a new facility in a tree-filled watershed would be less because no money would be spent trying to control the trees. However, the tree-filled watershed would yield 30 percent less water. In the final analysis, the ecologists found that the unit cost of water would be 14 percent lower from a facility developed in a treeless fynbos watershed, despite the ongoing cost of keeping trees out.

Maximizing water yield per dollar is critical, the researchers pointed out, because there aren't many appropriate sites left for new dams and water supply facilities. Fynbos watersheds already produce two-thirds of the water for the western Cape, including the rapidly growing cities of Cape Town and Port Elizabeth, as well as irrigated fields of wheat, wine grapes, and other fruit that generate hundreds of millions of dollars in exports each year. The only other options for new water supplies—reuse of sewage effluent or desalination of seawater—would cost two to seven times as much as developing the natural catchments.

The three noted there's another reason why paying to maintain a functional fynbos ecosystem makes good economic sense. The harvest and export of fynbos flowers, such as various species of proteas, already brings in $19 million a year. Rooibos tea harvested from another species brings in $2.1 million. The unique Cape flora, arguably the most biologically diverse of the earth's six recognized plant kingdoms with its 8,574 species, is the focus of a developing tourism market in South Africa.[22]

Ecologists and economists have not yet begun to evaluate the economic consequences of incremental losses, the slow erosion of services, and the weakening of systems that come with elimination of species. The fynbos, for example, has already lost 36 of its plant species to extinction, and another 618 are threatened.[23] Each loss, considered in isolation, may seem insignificant, yet the creeping attrition of populations and species leads to a biotic impoverishment and instability that must eventually compromise ecosystem services. At what point will an impoverished system reach some critical threshold beyond which further losses will cause it to falter?

Irreversible decisions about what to save and what to let go, based on today's valuations, would be plagued by too many uncertainties—what Bryan Norton calls the "oops" of chagrin and "boggles" of ignorance that fuzz any calculations about the dollar value of biodiversity. Norton likens the prospect of having to put a price on biodiversity right now to waking up in a hospital after a horrible accident, hooked up to a complicated life-support system and surrounded by cash-strapped, technologically ignorant hospital administrators. They are debating which bits of the life-support system they might purloin for a yard sale without fatally damaging its workings.[24] Many of our current decisions to "sell off" bits and pieces of our ecological life supports have a similar reckless quality to them.

Until more is known, human societies should be guided by what economists call the "safe minimum standard," treating each species as an irreplaceable resource that should be preserved for future generations unless the costs of doing so prove to be intolerably—not just inconveniently—high.

Summing Up the Work of Nature

Ecologists have only recently begun to investigate the functional role of biodiversity at all levels, from genes to species, communities, ecosystems, and landscapes. Although the field is young, clearly organisms have profound effects on the ecological processes that supply human beings with food, water, energy, clean air, and other services. It is not just numbers of species but their identity, locations, and interactions that are key to the workings of the earth's life-support systems. Let's recap some of the general patterns detailed throughout *The Work of Nature*.

The Keystone Club: Who's Important

Ecologists are still struggling to find guiding principles that might help identify which species are most critical to the health and persistence of communities and the functioning of larger-scale ecosystems. Still, they do know enough to offer some general guidelines:

- Even though identifying keystone or critical species has proven enormously difficult, more of these talented and influential organisms are at work in natural communities than was once believed.

- Adding or eliminating an abundant or dominant species, one whose sheer physical presence helps to define the community, changes the character of the assemblage and likely its functioning as well.

- Most communities contain some level of redundancy and overlap, usually in the form of rare or minor species that take over and compensate when others are lost. Until these backup workers can be identified with some confidence, rare species should never be viewed as expendable.

- Even in richly diverse communities, the complex web of interactions among species means that changes in one can snowball to affect others and may alter operations at the ecosystem level as well. In already-sparse communities, the impact will be still greater.

Community Ties

The immediate impact of eliminating native species or introducing exotics usually sets off a cascade of secondary effects that may take decades or centuries to make their marks. This cascade can disturb the community web by disrupting one or more classic ecological interactions: parasitism; pathogenicity; competition; mutualisms, such as pollination or seed dispersal; and consumption, either by grazing or predation.

By definition, changes in keystones or dominant species will cause a major shakeup in the web of community interdependencies. More subtle species losses can also create ripple effects that indirectly alter such services as pollination, natural pest control, and checks on disease agents and the organisms that vector them.

Water: The Essence of Life

Plant diversity strongly influences the fate of rainfall that strikes the land. On average, two-thirds of the water is cycled directly back to the atmosphere by plants, evaporating quickly from wet leaves and branches or being taken up and transpired by growing plants. The type of plant cover profoundly influences processes such as water yields and distribution, nutrient leaching and erosion, waterlogging and saliniza- tion of soils, and desertification of drylands. In addition, plants and plant debris, as well as microbes and animals ranging from beavers to fish exert strong impacts on water quality and the health and produc- tivity of aquatic habitats.

The Vitality of the Soil

The soil is a system maintained by a partnership between aboveground and belowground organisms. The vitality and character of the soil com- munity—agents of decay and renewal ranging from bacteria, fungi, and nematodes to earthworms—determine to a large extent what grows aboveground. The plant community in turn shelters and protects the soil from weathering and erosion and nurtures the belowground system with its carbon-rich debris and the sugars it yields up from growing roots. Disrupting the synchrony between these systems can affect the productivity of crops and wildlands and in turn the other services that a healthy aboveground community supports.

Of Plants and Productivity

Beyond some relatively low threshold number, short-term productivity does not seem to benefit from the addition of more plant species. Thus, in the near future, even increasingly impoverished ecosystems may re- main lush and productive. Even so, increasing evidence suggests that greater diversity provides for more robust and steady productivity over the long term. That's because a rich array of plant species insures that when drought or other environmental stresses cause some to falter or die back, other species with different tolerances will be available to compensate. With a growing human population that already cultivates a third of the earth's land surface and co-opts 40 percent of the potential productivity of all earth systems, chancing the wild swings in productivity that come with low-diversity systems is simply too risky. In both agricultural and natural systems, the focus should be on sustainable productivity rather than maximal short-term productivity.

The Power to Shape the Land

Climate and geology cannot be awarded sole credit for determining the layout of forests, grasslands, deserts, and other communities. Both plants and animals powerfully shape the living landscape, altering the complexity and variety of habitats, the architecture of plant communities, and even the diversity of plants and animals. The strong players, besides humans, include

- animals ranging from elephants and moose to bears and gophers whose feeding habits or earth-moving talents alter the landscape;
- organisms such as corals, kelps, and forest trees whose sheer physical presence defines the character of the landscape or seascape;
- trees, grasses, and shrubs that control the intensity and frequency of wildfires; and
- plants that array themselves into patterns that control the distribution of resources across the landscape.

Eliminating such organisms or disrupting their work can mean not just an emptier landscape but one that looks and functions differently and may not provide the services humans expect from woods, rangelands, or other natural systems.

Climate and Atmosphere

Terrestrial plant communities influence climate processes both directly and indirectly. First, vegetation generates surface conditions, such as roughness and albedo, at the boundary between the earth and air and determines evapotranspiration rates, thus influencing regional rainfall, temperature patterns, and air currents. Second, both plants and soils serve as sources and sinks for greenhouse gases, such as carbon dioxide, methane, and nitrous oxide, that change the heat-trapping properties of the atmosphere, indirectly influencing the global climate. Major realignments of the earth's vegetation zones, driven either by human land-use changes, such as deforestation, or by a warming climate, affect both the surface properties of the earth and the gas exchange between biosphere and atmosphere. These altered plant services then drive further changes in atmosphere and climate in a continuous push-pull between living things and the atmosphere. Even without the warming effect of greenhouse gases, human-driven buildup of carbon dioxide in the atmosphere will influence plant growth and alter key physiological processes, such as transpiration, that also have a profound effect on regional temperature and rainfall.

Conservation as Self-Preservation

These findings are not conclusions but starting points, the outlines of an emerging science. So little is known yet about how many and which species are present in most communities on the earth that it remains impossible to make firm pronouncements about the operational consequences of their loss. In some areas our ignorance is truly vast, from the biological riches and ecological workings of deep-sea communities far from view to the operations of soil communities in our own neighborhoods.

Despite this great uncertainty, policymakers tend to behave as though the survival of most non-human organisms is an amenity, one that future generations of humans can live without. Everything scientists are learning about the earth's life-support processes argues against this view. It's time to complement our sense of obligation as stewards of the earth with a somewhat humbler sense of self-preservation, to acknowledge that despite our increasing estrangement from nature, even urban societies are profoundly dependent on it. Self-preservation is no substitute for ethics, but it's a strong companion, less easily brushed aside in the hubbub of business as usual. Our interest in our own survival may also be the only thing people everywhere can agree on, and if it compels human societies to preserve the biological riches that nurture life as we know it, that's the best outcome we can ask.

Since scientists know so little about which organisms are critical to maintaining vital ecosystem services, the most prudent course for a survival-minded human species would be to exercise caution and work to preserve all of them. That's a social and economic choice many of us would prefer. Right now, the natural level of diversity is the best proxy scientists have for healthy functioning, and all species losses are warnings of a potential malfunction.

Unfortunately, it may not be feasible to prevent all further losses of biodiversity, even with the best intentions. Only 3 percent of the global land surface is set aside in parks and protected areas. More than 95 percent is already under direct human influence, whether plowed, paved, and managed intensively, or sparsely occupied by rural or indigenous peoples.[25] Seventy percent of the globe is covered by oceans and seas, yet only one-quarter of 1 percent of these are formally protected from exploitation and degradation. Nearly all of those protected waters are on the continental shelves rather than the continental slopes and deep seas that cover two-thirds of the earth.[26] With the human population increasing exponentially, our species is hardly likely to take up less space or exploit fewer resources in the future. Now, more than ever, we need to learn how to use lightly and sustainably the natural systems that survive in our midst, from swamps, coastal waters, savannas, and tropical forests to hedgerows and remnant woodlands along urban streams. The more we plunder them, the more likely we are to lower the earth's human carrying capacity—that is, its ability to support *Homo sapiens*.

Making intelligent decisions about what to save and what to let go is called triage, and it's an unpopular strategy among conservationists. Yet, by default, human societies are already choosing with their limited conservation dollars what they are willing to live without. What's needed now is a way to choose with greater wisdom. As Norman Myers writes, "We must come to grips with the facts: having goofed at playing Noah, we now find ourselves playing God."[27]

If we can't save everything, we can at least hope to better understand how the systems that supply us with food, water, energy, and other services will change. By trying to understand which biological resources provide necessary ecological services, and how natural systems might remain resilient in the face of global changes, we may eventually come to know which species and ecosystems are most critical to nurture us into the future.

Notes

Foreword

1. G. Daily, P. Ehrlich, and N. Haddad, 1993. "Double keystone bird in a keystone species complex," *Proceedings of the National Academy of Sciences* 90: 592–594.
2. M. Soulé (ed.), 1987. "A so-called 'minimum viable population'," *Viable Populations for Conservation,* Cambridge University Press, Cambridge.
3. P. Ehrlich and A. Ehrlich, 1996. *Betrayal of Science and Reason: How Anti-Environmental Rhetoric Threatens Our Future.* Island Press, Washington, D.C.
4. G. Easterbrook, 1995. *A Moment on the Earth,* Viking, New York, p. 9.
5. J. Holdren, 1991. "Population and the energy problem," *Population and Environment* 12: 231–255. Here, I am using per capita energy use as a surrogate for per capita environmental impact (consumption per person times a factor that measures the impact of the technologies used to service that consumption).

Chapter One

1. P. R. Ehrlich and A. H. Ehrlich, 1992. "The value of biodiversity," *Ambio* 21(3): 219–226.
2. P. R. Ehrlich and H. A. Mooney, 1983. "Extinction, substitution, and ecosystem services," *Bioscience* 33: 248–254.
3. H. T. Odum, 1971. *Environment, Power, and Society,* Wiley-Interscience, New York.

4. E. P. Odum, 1971. *Fundamentals of Ecology,* 3rd ed., W. B. Saunders Co., Philadelphia.

5. Technical volumes from the SCOPE project include:

E.-D. Schulze and H. A. Mooney (eds.), 1993. *Biodiversity and Ecosystem Function,* Springer-Verlag, Berlin.

H. A. Mooney, J. Lubchenco, R. Dirzo, and O. E. Sala (eds.), 1995. "Biodiversity and Ecosystem Functioning," in United Nations Environment Program, *Global Biodiversity Assessment,* Cambridge University Press, Cambridge, pp. 275–452.

H. A. Mooney, J. H. Cushman, E. Medina, O. E. Sala, and E.-D. Schulze (eds.), 1996. *Functional Roles of Biodiversity: A Global Perspective,* John Wiley & Sons, Chichester, U.K..

G. W. Davis and D. M. Richardson (eds.), 1995. *Mediterranean-Type Ecosystems: The Function of Biodiversity,* Springer-Verlag, Berlin.

R. J. Hobbs (ed.), 1992. *Biodiversity in Mediterranean Ecosystems of Australia,* Surrey Beatty and Sons, Chipping Norton, Australia.

F. S. Chapin III and C. Korner (eds.), 1995. *Arctic and Alpine Biodiversity: Patterns, Causes and Ecosystem Consequences,* Springer-Verlag, Berlin.

O. T. Solbrig, E. Medina, and J. F. Silva (eds.), 1996. *Biodiversity and Savanna Ecosystem Processes: A Global Perspective,* Springer-Verlag, Berlin.

G. H. Orians, R. Dirzo, and J. H. Cushman (eds.), 1996. *Biodiversity and Ecosystem Processes in Tropical Forests,* Springer-Verlag, Berlin.

P. M. Vitousek, L. L. Loope, and H. Adersen (eds.), 1995. *Islands: Biological Diversity and Ecosystem Function,* Springer-Verlag, Berlin.

6. N. Myers, 1993. "Biodiversity and the precautionary principle," *Ambio* 22(2–3): 74–79.

7. Ehrlich and Ehrlich, 1992.

8. P. R. Ehrlich and E. O. Wilson, 1991. "Biodiversity studies: Science and policy," *Science* 253: 758–762.

9. E. O. Wilson, 1989. "Threats to biodiversity," *Scientific American,* September, pp. 108–116.

10. United Nations Environment Program, 1995. *Global Biodiversity Assessment: Summary for Policy-Makers,* Cambridge University Press, Cambridge.

11. Ehrlich and Wilson, 1991.

12. S. L. Pimm, G. J. Russell, J. L. Gittleman, and T. M Brooks, 1995. "The future of biodiversity," *Science* 269: 347–350.

13. P. R. Ehrlich and G. C. Daily, 1993. "Population extinction and saving biodiversity," *Ambio* 22(2–3): 64–68.

14. P. R. Ehrlich. 1991, "Population diversity and the future of ecosystems," *Science* 254: 175.

15. E. Wilken, 1995. "Urbanization spreading," in L. R. Brown et al. (eds.), *Vital Signs 1995,* Worldwatch Institute/W.W. Norton, New York, pp. 100–101.

Chapter Two

1. M. E. Power et al., 1996. "Challenges in the quest for keystones," *Bioscience* 46: 609–620.
2. H. A. Mooney, J. Lubchenco, R. Dirzo, and O. E. Sala (eds.), 1995. "Biodiversity and Ecosystem Functioning," in United Nations Environment Program, *Global Biodiversity Assessment,* Cambridge University Press, Cambridge, pp. 275–452.
3. P. R. Ehrlich, 1993. "Biodiversity and ecosystem function: Need we know more?" Foreword, in E.-D. Schulze and H. A. Mooney (eds.), *Biodiversity and Ecosystem Function,* Springer-Verlag, Berlin, pp. vii–xi.
4. P. R. Ehrlich and A. H. Ehrlich, 1981. *Extinction: The Causes and Consequences of the Disappearance of Species,* Random House, New York, pp. xii–xiii.
5. B. H. Walker, 1992. "Biodiversity and ecological redundancy," *Conservation Biology* 6: 18–23.
6. Mooney et al., 1995.
7. F. I. Woodward, 1993. "How many species are required for a functional ecosystem," in E.-D. Schulze and H. A. Mooney (eds.), *Biodiversity and Ecosystem Function,* Springer-Verlag, Berlin, pp. 271–291.
8. E. I. Friedmann, M. Hua, and R. Ocampo-Friedmann, 1988. "Cryptoendolithic lichen and cyanobacterial communities of the Ross Desert, Antarctica," *Polarforschung* 58: 251–259.
9. S. J. Wright, 1996. "Plant species diversity and ecosystem functioning in tropical forests," in G. H. Orians, R. Dirzo, and J. H. Cushman (eds.), *Biodiversity and Ecosystem Processes in Tropical Forests,* Springer-Verlag, Berlin, pp. 11–31.
10. Y. Iwasa, K. Sato, M. Kakita, and T. Kubo, 1993. "Modelling biodiversity: Latitudinal gradient of forest species diversity," in E.-D. Schulze and H. A. Mooney (eds.), *Biodiversity and Ecosystem Function,* Springer-Verlag, Berlin, pp. 433–451.
11. Woodward, 1993.
12. P. M. Vitousek and D. U. Hooper, 1993. "Biological diversity and terrestrial ecosystem biogeochemistry," in E.-D. Schulze and H. A. Mooney (eds.), *Biodiversity and Ecosystem Function,* Springer-Verlag, Berlin, pp. 3–14.
13. J. J. Ewel, M. J. Mazzarino, and C. W. Berish, 1991. "Tropical soil fertility changes under monocultures and successional communities of different structure," *Ecological Applications* 1(3): 289–302.
14. Wright, 1996.
15. Vitousek and Hooper, 1993.
16. Mooney et al., 1995.
17. S. J. McNaughton, 1988. "Diversity and stability," *Nature* 333: 204–205.
18. D. Tilman, 1996. "Biodiversity: Population versus ecosystem stability," *Ecology* 77: 350–363.

19. B. A. Menge, J. Lubchenco, L. R. Ashkenas, and F. Ramsey, 1986. "Experimental separation of effects of consumers on sessile prey in the low zone of a rocky shore in the Bay of Panama: Direct and indirect consequences of food web complexity," *Journal of Experimental Marine Biology and Ecology* 100: 225–269.

20. T. M. Frost, S. R. Carpenter, A. R. Ives, and T. K. Kratz, 1995. "Species compensation and complementarity in ecosystem function," in C. G. Jones and J. H. Lawton (eds.), *Linking Species and Ecosystems,* Chapman and Hall, New York, pp. 224–239.

21. D. W. Schindler, 1995. "Linking species and communities to ecosystem management: A perspective from the experimental lakes experience," in C. G. Jones and J. H. Lawton (eds.), *Linking Species and Ecosystems,* Chapman and Hall, New York, pp. 313–325.

22. Frost et al., 1995.

23. S. R. Carpenter, T. M. Frost, J. F. Kitchell, and T. K. Kratz, 1993. "Species dynamics and global environmental change: A perspective from ecosystem experiments," in P. M. Kareiva, J. G. Kingsolver, and R. B. Huey (eds.), *Biotic Interactions and Global Change,* Sinauer Associates, Sunderland, Massachusetts, pp. 267–279.

24. Mooney et al., 1995.

25. H. H. Shugart, 1984. *A Theory of Forest Dynamics,* Springer-Verlag, Berlin.

26. R. T. Paine, 1966. "Food web complexity and species diversity," *American Naturalist* 100: 65–75.

27. R. T. Paine, 1969. "A note on trophic complexity and community stability," *American Naturalist* 103: 91–93.

28. J. A. Estes and J. F. Palmisano, 1974. "Sea otters: Their role in structuring nearshore communities," *Science* 185: 1058–1060.

29. R. G. Kvitek, J. S. Oliver, A. R. DeGange, and B. S. Anderson, 1992. "Changes in Alaskan soft-bottom prey communities along a gradient in sea otter predation," *Ecology* 73: 413–428.

30. W. J. Bond, 1993. "Keystone species," in E.-D. Schulze and H. A. Mooney (eds.), *Biodiversity and Ecosystem Function,* Springer-Verlag, Berlin, pp. 237–253.

31. J. H. Brown and E. J. Heske, 1990. "Control of a desert–grassland transition by a keystone rodent guild," *Science* 250: 1705–1707.

32. C. G. Jones, J. H. Lawton, and M. Shachak, 1994. "Organisms as ecosystem engineers," *Oikos* 69: 373–386.

33. J. H. Lawton and C. G. Jones, 1995. "Linking species and ecosystems: Organisms as ecosystem engineers," in C. G. Jones and J. H. Lawton (eds.), *Linking Species and Ecosystems,* Chapman and Hall, New York, pp. 141–150.

34. J. M. Klopatek and W. D. Stock, 1994. "Partitioning of nutrients in *Acanthosicyos horridus,* a keystone endemic species in the Namib Desert," *Journal of Arid Environments* 26: 233–240.

35. W. J. Bond, 1993.

36. R. J. Naiman, C. A. Johnston, and J. C. Kelley, 1988. "Alteration of North American streams by beaver," *BioScience* 38: 753–763.

37. M. Shachak, C. G. Jones, and Y. Granot, 1987. "Herbivory in rocks and the weathering of a desert," *Science* 236: 1098–1099.

38. C. G. Jones and M. Shachak, 1990. "Fertilization of the desert soil by rock-eating snails," *Nature* 346: 839–841.

39. C. G. Jones et al., 1994.

40. T. J. Smith III, K. G. Boto, S. D. Frusher, and R. L. Giddins, 1991. "Keystone species and mangrove forest dynamics: The influence of burrowing by crabs on soil nutrient status and forest productivity," *Estuarine, Coastal and Shelf Science* 33: 419–432.

41. W. J. Bond, 1993.

42. L. S. Mills, M. E. Soulé, and D. F. Doak, 1993. "The keystone-species concept in ecology and conservation," *BioScience* 43: 219–224.

43. R. T. Paine, 1995. "A conversation on refining the concept of keystone species," *Conservation Biology* 9: 962–964.

44. M. E. Power and L. S. Mills, 1995, "The keystone cops meet in Hilo," *Trends in Ecology and Evolution* 10: 182–184.

45. M. E. Power et al., 1996.

46. P. R. Ehrlich, 1985. "Extinctions and ecosystem functions," in R. J. Hoage (ed.), *Animal Extinctions: What Everyone Should Know,* Smithsonian Institution Press, Washington, D.C., pp. 159–173.

47. J. C. Castilla and L. R. Duran, 1985. "Human exclusion from the rocky intertidal zone of central Chile: The effects of *Concholepas concholepas* (Gastropoda)," *Oikos* 45: 391–399.

48. L. R. Duran and J. C. Castilla, 1989. "Variation and persistence of the middle rocky intertidal community of central Chile, with and without human harvesting," *Marine Biology* 103: 555–562.

49. M. E. Power et al., 1996.

50. B. A. Menge, E. L. Berlow, C. A. Blanchette, S. A. Navarrete, and S. B. Yamada, 1994. "The keystone species concept: Variation in interaction strength in a rocky intertidal habitat," *Ecological Monographs* 64: 249–286.

51. A. Barkai and C. McQuaid, 1988. "Predator–prey role reversal in a marine benthic ecosystem," *Science* 242: 62–64.

52. P. H. Raven, 1976. "Ethics and attitudes," in J. Simmons et al. (eds.), *Conservation of Threatened Plants,* Plenum Publishing, New York, pp. 155–181.

53. L. E. Gilbert, 1980. "Food web organization and the conservation of neotropical diversity," in M. E. Soulé and B. A. Wilcox (eds.), *Conservation Biology: An Evolutionary-Ecological Perspective,* Sinauer Associates, Sunderland, Massachusetts, pp. 11–33.

Chapter Three

1. C. N. Spencer, B. R. McClelland, and J. A. Stanford, 1991. "Shrimp stocking, salmon collapse, and eagle displacement," *BioScience* 41: 14–21.

2. J. Boswall, 1986. "Notes on the current status of ornithology in the People's Republic of China," *Forktail* 2: 43–51.

3. N. Myers, 1979. "China's wildlife," *National Parks and Conservation Magazine,* September, pp. 10–15.

4. W. I. Aron and S. H. Smith, 1971. "Ship canals and aquatic ecosystems," *Science* 174: 13–20.

5. D. H. Janzen, 1974. "The deflowering of Central America," *Natural History* 83(4): 49–53.

6. M. A. Aizen and P. Feinsinger, 1994. "Forest fragmentation, pollination, and plant reproduction in a Chaco dry forest, Argentina," *Ecology* 75: 330–351.

7. D. J. Howell and B. S. Roth, 1981. "Sexual reproduction in agaves: The benefits of bats; the cost of semelparous advertising," *Ecology* 62: 1–7.

8. J. H. Cushman, 1995. "Ecosystem-level consequences of species additions and deletions on islands," in P. M. Vitousek, L. L. Loope, and H. Adersen (eds.), *Islands: Biological Diversity and Ecosystem Function,* Springer-Verlag, Berlin, pp. 135–147.

9. P. A. Cox, T. Elmqvist, E. D. Pierson, and W. E. Rainey, 1991. "Flying foxes as strong interactors in South Pacific island ecosystems: A conservation hypothesis," *Conservation Biology* 5: 448–454.

10. O. Jennersten, 1988. "Pollination in *Dianthus deltoides* (Caryophyllaceae): Effects of habitat fragmentation on visitation and seed set," *Conservation Biology* 2: 359–366.

11. W. J. Bond, 1994. "Do mutualisms matter? Assessing the impact of pollinator and disperser disruption on plant extinction," *Philosophical Transactions of the Royal Society of London* B 344: 83–90.

12. L. E. Gilbert, 1980. "Food web organization and the conservation of neotropical diversity," in M. E. Soulé and B. A. Wilcox (eds.), *Conservation Biology: An Evolutionary-Ecological Perspective,* Sinauer Associates, Sunderland, Massachusetts, pp. 11–33.

13. R. J. Hobbs et al., 1995. "Function of biodiversity in the Mediterranean-type ecosystems of southwestern Australia," in G. W. Davis and D. M. Richardson (eds.), *Mediterranean-Type Ecosystems: The Function of Biodiversity,* Springer-Verlag, Berlin, pp. 233–284.

14. H. F. Howe, 1984. "Implications of seed dispersal by animals for tropical reserve management," *Biological Conservation* 30: 261–281.

15. J. Terborgh, 1986. "Keystone plant resources in the tropical forest," in M. E. Soulé (ed.), *Conservation Biology: The Science of Scarcity and Diversity,* Sinauer Associates, Sunderland, Massachusetts, pp. 330–344.

16. G. H. Kattan, 1992. "Rarity and vulnerability: The birds of the Cordillera Central of Colombia," *Conservation Biology* 6: 64–70.

17. J. Terborgh, 1989. *Where Have All the Birds Gone?* Princeton University Press, Princeton, New Jersey.

18. R. J. Marquis and C. J. Whelan, 1994. "Insectivorous birds increase growth of white oak through consumption of leaf-chewing insects," *Ecology* 75: 2007–2014.

19. H. Youth, 1994. "Flying into trouble," *World Watch,* January/February, pp. 10–19.

20. D. H. Janzen and P. S. Martin, 1982. "Neotropical anachronisms: The fruits the gomphotheres ate," *Science* 215: 19–27.

21. J. Terborgh, 1988. "The big things that run the world—A sequel to E. O. Wilson," *Conservation Biology* 2: 402–403.

22. J. Terborgh, 1992. *Diversity and the Tropical Rain Forest,* Scientific American Library, New York.

23. R. Dirzo and A. Miranda, 1991. "Altered patterns of herbivory and diversity in the forest understory: A case study of the possible consequences of contemporary defaunation," in P. W. Price et al. (eds.), *Plant–Animal Interactions: Evolutionary Ecology in Tropical and Temperate Regions,* John Wiley & Sons, New York.

24. R. Dirzo and A. Miranda, 1990. "Contemporary neotropical defaunation and forest structure, function, and diversity—A sequel to John Terborgh," *Conservation Biology* 4: 444–447.

25. J. Lubchenco, 1978. "Plant species diversity in a marine intertidal community: Importance of herbivore food preference and algal competitive abilities," *American Naturalist* 112: 23–39.

26. N. Huntly, 1991. "Herbivores and the dynamics of communities and ecosystems," *Annual Review of Ecology and Systematics* 22: 477–503.

27. R. L. Jefferies and J. P. Bryant, 1995. "The plant–vertebrate herbivore interface in Arctic ecosystems," in F. S. Chapin III and C. Korner (eds.), *Arctic and Alpine Biodiversity: Patterns, Causes and Ecosystem Consequences,* Springer-Verlag, Berlin, pp. 271–281.

28. C. Bright, 1996. "Understanding the threat of bioinvasions," in L. R. Brown et al., (eds.), *State of the World,* WorldWatch Institute/W.W. Norton & Co., New York, pp. 95–113.

29. E. Culotta, 1991. "Biological immigrants under fire," *Science* 254: 1444–1447.

30. Spencer et al., 1991.

31. Culotta, 1991.

32. P. M. Vitousek, 1990. "Biological invasions and ecosystem processes: Towards an integration of population biology and ecosystem studies," *Oikos* 57: 7–13.

33. Culotta, 1991.

34. Spencer et al., 1991.

35. J. D. Varley and P. Schullery, 1995. *The Yellowstone Lake Crisis: Confronting a Lake Trout Invasion,* U.S. National Park Service, Yellowstone National Park, Wyoming.

36. U.S. Congress, Office of Technology Assessment, 1993. *Harmful Non-indigenous Species in the United States,* OTA-F-565, U.S. Government Printing Office, Washington, D.C.

37. Varley and Schullery, 1995.

38. Aron and Smith, 1971.

39. U.S. Congress, Office of Technology Assessment, 1993.

40. J. J. Burdon, 1993. "The role of parasites in plant populations and communities," in E.-D. Schulze and H. A. Mooney (eds.), *Biodiversity and Ecosystem Function,* Springer-Verlag, Berlin, pp. 165–179.

41. J. D. Castello, D. J. Leopold, and P. J. Smallidge, 1995. "Pathogens, patterns, and processes in forest ecosystems," *BioScience* 45: 16–24.

42. P. A. Matson and R. D. Boone, 1984. "Natural disturbance and nitrogen mineralization: Wave-form dieback of mountain hemlock in the Oregon Cascades," *Ecology* 65: 1511–1516.

43. E. L. Simms, 1996. "The evolutionary genetics of plant-pathogen systems," *BioScience* 46: 136–145.

44. G. S. Gilbert and S. P. Hubbell, 1996. "Plant diseases and the conservation of tropical forests," *BioScience* 46: 98–106.

45. R. T. Wills, 1992. "The ecological impact of *Phytophthora cinnamomi* in the Stirling Range National Park, Western Australia," *Australian Journal of Ecology* 17: 145–159.

46. F. D. Podger and M. J. Brown, 1989. "Vegetation damage caused by *Phytophthora cinnamomi* on disturbed sites in temperate rainforest in western Tasmania," *Australian Journal of Botany* 37: 443–480.

47. Castello et al., 1995.

48. K. C. Kendall and S. F. Arno, 1990. "Whitebark pine—an important but endangered wildlife resource," *Proceedings—Symposium on Whitebark Pine Ecosystems: Ecology and Management of a High-Mountain Resource,* U.S. Department of Agriculture, Forest Service, General Technical Report INT-270.

49. K. C. Kendall, 1995. "Whitebark pine: Ecosystem in peril," in E. T. LaRoe, et al. (eds.), *Our Living Resources: A Report to the Nation on the Distribution, Abundance, and Health of U.S. Plants, Animals, and Ecosystems,* U.S. Department of the Interior, National Biological Service, Washington, D.C.

50. S. J. McNaughton, 1989. "Ecosystems and conservation in the twenty-first century," in D. Western and M. C. Pearl (eds.), *Conservation for the Twenty-first Century,* Oxford University Press, New York, pp. 117–118.

51. C. R. Carroll, 1990. "The interface between natural areas and agroecosystems," in C. R. Carroll, J. H. Vandermeer, and P. Rosset (eds.), *Agroecology,* McGraw-Hill, New York, pp. 365–383.

52. Bright, 1996.

53. V. Morell, 1996. "New virus variant killed Serengeti cats," *Science* 271: 596.

54. E. Royte, 1995. "On the brink: Hawaii's vanishing species," *National Geographic,* September, pp. 2–37.

55. M. E. Watanabe, 1994. "Pollination worries rise as honey bees decline," *Science* 265: 1170.

56. Bright, 1996.

57. M. A. Hoy, 1990. "The importance of biological control in U.S. agriculture," *Journal of Sustainable Agriculture* 1: 59–79.

58. S. W. T. Batra, 1981. "Biological control in agroecosystems," *Science* 215: 134–139.

59. J. K. Waage, 1991. "Biodiversity as a resource for biological control," in D. L. Hawksworth (ed.), *The Biodiversity of Microorganisms and Invertebrates: Its Role in Sustainable Agriculture,* CAB International, Wallingford, U.K., pp. 149–163.

60. M. W. Rosegrant and P. L. Pingall, 1991. "Sustaining rice productivity growth in Asia: A policy perspective," *IRRI Social Science Division Papers,* No. 91-01, International Rice Research Institute, Manila, Philippines.

61. G. Gardner, 1996. "Preserving agricultural resources," in L. R. Brown et al. (eds.), *State of the World,* WorldWatch Institute/W.W. Norton & Co., New York, pp. 78–94.

62. S. H. Verhovek, 1996. "In Texas, an attempt to swat an old pest stirs a revolt," *The New York Times,* January 24, 1996.

63. Castello et al., 1995.

64. E.-D. Schulze and P. Gerstberger, 1993. "Functional aspects of landscape diversity: A Bavarian example," in E.-D. Schulze and H. A. Mooney (eds.), *Biodiversity and Ecosystem Function,* Springer-Verlag, Berlin, pp. 453–466.

65. D. Pimentel, 1986. "Biological invasions of plants and animals in agriculture and forestry," in H. A. Mooney and J. A. Drake (eds.), *Ecology of Biological Invasions of North America and Hawaii,* Springer-Verlag, New York, pp. 149–162.

66. J. L. Ruesink, I. M. Parker, M. J. Groom, and P. M. Kareiva, 1995. "Reducing the risks of non-indigenous species introductions," *BioScience* 45: 465–477.

67. Culotta, 1991.

68. Bright, 1996.

69. A. E. Platt, 1996. "Confronting infectious diseases," in L. R. Brown et al. (eds.), *State of the World,* WorldWatch Institute/W.W. Norton & Co., New York, pp. 114–132.

70. J. Sterngold, 1996. "Japan's cedar forests are man-made disaster," *The New York Times,* January 17, 1995.

71. R. S. Ostfeld, C. G. Jones, and J. O. Wolff, 1996. "Of mice and mast: Ecological connections in eastern deciduous forests," *BioScience* 46: 323–330.

72. L. A. Real, 1996. "Sustainability and the ecology of infectious disease," *BioScience* 46: 88–97.

73. P. A. Roger, K. L. Heong, and P. S. Teng, 1991. "Biodiversity and sustainability of wetland rice production: Role and potential of microorganisms and invertebrates," in D. L. Hawksworth (ed.), *The Biodiversity of Microorganisms and Invertebrates: Its Role in Sustainable Agriculture,* CAB International, Wallingford, U.K., pp. 117–136.

74. Platt, 1996.

75. T. J. Smayda, 1990. "Novel and nuisance phytoplankton blooms in the sea: Evidence for a global epidemic," in E. Graneli et al. (eds.), *Toxic Marine Phytoplankton,* Elsevier Science Publishing Co., Amsterdam, pp. 29–40.

76. T. J. Smayda and A. W. White, 1990. "Has there been a global expansion of algal blooms? If so, is there a connection with human activities?" in E. Graneli et al. (eds.), *Toxic Marine Phytoplankton,* Elsevier Science Publishing Co., Amsterdam, pp. 516–517.

Chapter Four

1. P. P. Rogers, 1985. "Fresh water," in R. Repetto (ed.), *The Global Possible: Resources, Development, and the New Century,* Yale University Press, New Haven and London, pp. 255–298.
2. G. Gardner, 1995. "Water tables falling," in L. R. Brown et al. (eds.), *Vital Signs 1995,* Worldwatch Institute/W.W. Norton, New York, pp. 122–123.
3. M. Dynesius and C. Nilsson, 1994. "Fragmentation and flow regulation of river systems in the northern third of the world," *Science* 266: 753–762.
4. J. W. Maurits la Riviere, 1989. "Threats to the world's water," *Scientific American,* September, pp. 80–94.
5. Rogers, 1985.
6. R. J. Hobbs and D. A. Saunders (eds), 1993. *Reintegrating Fragmented Landscapes: Towards Sustainable Production and Nature Conservation,* Springer-Verlag, New York.
7. R. A. Nulsen, K. J. Bligh, I. N. Baxter, E. J. Solin, and D. H. Imrie, 1986. "The fate of rainfall in a mallee and heath vegetated catchment in southern Western Australia," *Australian Journal of Ecology* 11: 361–371.
8. S. Postel, 1994. "Irrigation expansion slowing," in L. R. Brown, et al. (eds.), *Vital Signs 1994,* Worldwatch Institute/W.W. Norton, New York, pp. 44–45.
9. Nulsen et al., 1986.
10. Hobbs et al., 1993.
11. G. C. Daily, 1995. "Restoring value to the world's degraded lands," *Science* 269: 350–354.
12. S. Williams, 1994. "Holding Back the Desert," *UNESCO Sources* 63: 23.
13. W. H. Schlesinger, J. F. Reynolds, G. L. Cunningham, L. F. Huenneke, W. M. Jarrell, R. A. Virginia, and W. G. Whitford, 1990. "Biological feedbacks in global desertification," *Science* 247: 1043–1048.
14. W. H. Schlesinger, J. A. Raikes, A. E. Hartley, and A. F. Cross, 1996. "On the spatial pattern of soil nutrients in desert ecosystems," *Ecology* 77: 364–374.
15. L. F. Huenneke and I. Noble, 1996. "Ecosystem function of biodiversity in arid ecosystems," in H. A. Mooney, J. H. Cushman, E. Medina, O. E. Sala, and E.-D. Schulze (eds.), *Functional Roles of Biodiversity: A Global Perspective,* John Wiley & Sons, Chichester, U.K., pp. 99–128.
16. O. T. Solbrig, E. Medina, and J. F. Silva, 1996. "Biodiversity and tropical savanna properties: A global view," in H. A. Mooney, J. H. Cushman, E. Medina, O. E. Sala, and E.-D. Schulze (eds.), *Functional Roles of Biodiversity: A Global Perspective,* John Wiley & Sons, Chichester, U.K., pp. 185–211.

17. W. T. Knoop and B. H. Walker, 1985. "Interactions of woody and herbaceous vegetation in a Southern African savanna," *Journal of Ecology* 73: 235–253.

18. Solbrig et al., 1996.

19. Schlesinger et al., 1990.

20. J. H. Patric, 1961. "The San Dimas large lysimeters," *Journal of Soil and Water Conservation* 16: 13–17.

21. R. M. Rice and G. T. Foggin III, 1971. "Effect of high intensity storms on soil slippage on mountainous watersheds in Southern California," *Water Resources Research* 7: 1485–1496.

22. W. T. Swank and J. E. Douglass, 1974. "Streamflow greatly reduced by converting deciduous hardwood stands to pine," *Science* 185: 857–859.

23. W. T. Swank and J. M. Vose, 1994. "Long-term hydrologic and stream chemistry responses of southern Appalachian catchments following conversion from mixed hardwoods to white pine," in R. Landolt (ed.), *Hydrologie kleiner Einzugsgebiete: Gedenkschrift Hans M. Keller,* Schweizerische Gesellschaft fur Hydrologie und Limnologie, Bern, pp. 164–172.

24. W. T. Swank, L. W. Swift Jr., and J. E. Douglass, 1988. "Streamflow changes associated with forest cutting, species conversions, and natural disturbances," in W. T. Swank and D. A. Crossley Jr. (eds.), *Forest Hydrology and Ecology at Coweeta,* Springer-Verlag, New York, pp. 297–312.

25. R. H. Waring and W. H. Schlesinger, 1985. *Forest Ecosystems: Concepts and Management,* Academic Press, Orlando, Florida.

26. G. M. Lovett, W. A. Reiners, and R. K Olson, 1982. "Cloud droplet deposition in subalpine balsam fir forests: Hydrological and chemical inputs," *Science* 218: 1303–1304.

27. H. W. Vogelmann, 1976. "Rain-making forests," *Natural History* 85 (March): 22–24.

28. B. W. van Wilgen, R. M. Cowling, and C. J. Burgers, 1996. "Valuation of ecosystem services: A case study from South African fynbos ecosystems," *BioScience* 46: 184–189.

29. W. L. Graf, 1978. "Fluvial adjustments to the spread of tamarisk in the Colorado Plateau region," *Geological Society of America Bulletin* 89: 1491–1501.

30. R. F. Noss and A. Y. Cooperrider, 1994. *Saving Nature's Legacy: Protecting and Restoring Biodiversity,* Island Press, Washington, D.C., p. 277.

31. L. R. Brown and J. E. Young, 1990. "Feeding the world in the Nineties," in L. R. Brown et al. (eds.), *State of the World,* WorldWatch Institute/W.W. Norton & Co., New York, pp. 59–78.

32. E. Wilken, 1995. "Soil erosion's toll continues," in L. R. Brown et al. (eds.), *Vital Signs 1995,* WorldWatch Institute/W.W. Norton & Co., New York, pp. 118–119.

33. J. Cherfas, 1990. "FAO Proposes a 'New' Plan for Feeding Africa," *Science* 250: 748–749.

34. J. Soulé, D. Carré, and W. Jackson, 1990. "Ecological impact of modern agri-

culture," in C. R. Carroll, J. H. Vandermeer, and P.M. Rosset (eds.), *Agroecology,* McGraw-Hill, New York, pp. 165–188.

35. J. M. Maass, C. F. Jordan, and J. Sarukhan, 1988. "Soil erosion and nutrient losses in seasonal tropical agroecosystems under various management techniques," *Journal of Applied Ecology* 25: 595–607.

36. Brown and Young, 1990.

37. T. Horton, 1993, "Chesapeake Bay: Hanging in the balance," *National Geographic,* June, pp. 2–35.

38. W. T. Peterjohn and D. L. Correll, 1984. "Nutrient dynamics in an agricultural watershed: Observations on the role of a riparian forest," *Ecology* 65: 1466–1475.

39. S. R. Carpenter, J. F. Kitchell, and J. R. Hodgson, 1985. "Cascading trophic interactions and lake productivity," *BioScience* 35: 634–639.

40. S. R. Carpenter and J. F. Kitchell, 1988. "Consumer control of lake productivity," *BioScience* 38: 764–769.

41. Y. Baskin, 1992. "Africa's troubled waters," *BioScience* 42: 476–481.

42. L. Kaufman, 1992. "Catastrophic change in species-rich freshwater ecosystems: The lessons of Lake Victoria," *BioScience* 42: 846–858.

43. M. M. Pollock, R. J. Naiman, H. E. Erickson, C. A. Johnston, J. Pastor, and G. Pinay, 1995. "Beavers as engineers: Influences on biotic and abiotic characteristics of drainage basins," in C. G. Jones and J. H. Lawton (eds.), *Linking Species and Ecosystems,* Chapman and Hall, New York, pp. 117–126.

44. R. J. Naiman, G. Pinay, C. A. Johnston, and J. Pastor, 1994. "Beaver influences on the long-term biogeochemical characteristics of boreal forest drainage networks," *Ecology* 75: 905–921.

45. R. J. Naiman, C. A. Johnston, and J. C. Kelley, 1988. "Alteration of North American streams by beaver," *BioScience* 38: 753–763.

46. C. A. Johnston and R. J. Naiman, 1990. "Aquatic patch creation in relation to beaver population trends," *Ecology* 71: 1617–1621.

47. Waring and Schlesinger, 1985.

48. C. Maser and J. R. Sedell, 1994. *From the Forest to the Sea: The Ecology of Wood in Streams, Rivers, Estuaries, and Oceans,* St. Lucie Press, Delray Beach, Florida.

49. S. V. Gregory, F. J. Swanson, W. A. McKee, and K. W. Cummins, 1991. "An ecosystem perspective of riparian zones," *BioScience* 41: 540–550.

50. R. E. Bilby and G. E. Likens, 1980. "Importance of organic debris dams in the structure and function of stream ecosystems," *Ecology* 61: 1107–1113.

51. W. J. Junk and K. Furch, 1985. "The physical and chemical properties of Amazonian waters and their relationships with the biota," in G. T. Prance and T. E. Lovejoy (eds), *Amazonia: Key Environments Series,* Pergamon Press, Oxford.

52. S. C. Snedaker, 1989. "Overview of ecology of mangroves and information needs for Florida Bay," *Bulletin of Marine Science* 44: 341–347.

53. R. R. Twilley, 1988. "Coupling of mangroves to the productivity of estuarine

and coastal waters," in B.O. Jansson (ed.), *Coastal-Offshore Ecosystem Interactions,* Springer-Verlag, Berlin, pp. 155–180.

54. R. R. Twilley, S. C. Snedaker, A. Y. Arancibia, and E. Medina, 1996. "Biodiversity and ecosystem processes in tropical estuaries: Perspectives from mangrove ecosystems," in H. A. Mooney, J. H. Cushman, E. Medina, O. E. Sala, and E.-D. Schulze (eds.), *Functional Roles of Biodiversity: A Global Perspective,* John Wiley & Sons, Chichester, U.K., pp. 327–370.

55. R. R. Twilley, A. Bodero, and D. Robadue, 1993. "Mangrove ecosystem biodiversity and conservation in Ecuador," in C. S. Potter, J. I. Cohen, and D. Janczewski (eds.), *Perspectives on Biodiversity: Case Studies of Genetic Resource Conservation and Development,* AAAS Press, Washington, D.C., pp. 105–127.

Chapter Five

1. D. D. Richter and D. Markewitz, 1995. "How deep is soil?" *BioScience* 45: 600–609.

2. D. G. Kaufman and C. M. Franz, 1993. *Biosphere 2000: Protecting Our Global Environment,* HarperCollins College Publishers, New York.

3. G. C. Daily, 1995. "Restoring value to the world's degraded lands," *Science* 269: 350–354.

4. D. A. Perry, J. G. Borchers, S. L. Borchers, and M. P. Amaranthus, 1990. "Species migrations and ecosystem stability during climate change: The belowground connection," *Conservation Biology* 4: 266–273.

5. H. A. Mooney, J. Lubchenco, R. Dirzo, and O. E. Sala (eds.), 1995. "Biodiversity and Ecosystem Functioning," in United Nations Environment Program, *Global Biodiversity Assessment,* Cambridge University Press, Cambridge, pp. 275–452.

6. Richter and Markewitz, 1995.

7. J. M. Anderson, 1988. "Spatiotemporal effects of invertebrates on soil processes," *Biology and Fertility of Soils* 6: 216–227.

8. P. Lashmar, 1994. "Invisible biodiversity," *UNESCO Sources 60,* July/August, p. 10.

9. O. Meyer, 1993. "Functional groups of microorganisms," in E.-D. Schulze and H.A. Mooney (eds.), *Biodiversity and Ecosystem Function,* Springer-Verlag, Berlin, pp. 67–96.

10. Anderson, 1988.

11. D. Boucher, 1990. "Beneficials in agricultural soils," in C. R. Carroll, J. H. Vandermeer, and P. Rosset (eds.), *Agroecology,* McGraw-Hill, New York, pp. 329–339.

12. R. E. Ingham, J. A. Trofymow, E. R. Ingham, and D. C. Coleman, 1985. "Interactions of bacteria, fungi, and their nematode grazers: Effects on nutrient cycling and plant growth," *Ecological Monographs* 55: 119–140.

13. H. A. Verhoef and L. Brussaard, 1990. "Decomposition and nitrogen mineralization in natural and agroecosystems: The contribution of soil animals," *Biogeochemistry* 11: 175–211.

14. Anderson, 1988.

15. A. D. Rovira, 1976. "Studies on soil fumigation—I: Effects on ammonium, nitrate, and phosphate in soil and on the growth, nutrition, and yield of wheat," *Soil Biology and Biochemistry* 8: 241–247.

16. E. H. Ridge, 1976. "Studies on soil fumigation—II: Effects on bacteria," *Soil Biology and Biochemistry* 8: 249–253.

17. E. R. Ingham, J. A. Trofymow, R. N. Ames, H. W. Hunt, C. R. Morley, J. C. Moore, and D. C. Coleman, 1986. "Trophic interactions and nitrogen cycling in a semi-arid grassland soil—II: System responses to the removal of different groups of soil microbes or fauna." *Journal of Applied Ecology* 23: 615–630.

18. P. O. Salonius, 1981. "Metabolic capabilities of forest soil microbial populations with reduced species diversity," *Soil Biology and Biochemistry* 13: 1–10.

19. Perry et al., 1990.

20. P. M. Vitousek, L. R. Walker, L. D. Whiteaker, D. Mueller-Dombois, and P. A. Matson, 1987. "Biological invasion by *Myrica faya* alters ecosystem development in Hawaii," *Science* 238: 802–804.

21. P. M. Vitousek and L. R. Walker, 1989. "Biological invasion by *Myrica faya* in Hawaii: Plant demography, nitrogen fixation, ecosystem effects," *Ecological Monographs* 59: 247–265.

22. D. Mueller-Dombois, 1990. "Impoverishment in Pacific island forests," in G. M. Woodwell (ed.), *The Earth in Transition,* Cambridge University Press, New York, pp. 199–210.

23. H. W. Hunt, E. R. Ingham, D. C. Coleman, E. T. Elliott, and C. P. P. Reid, 1988. "Nitrogen limitation of production and decomposition in prairie, mountain meadow, and pine forest," *Ecology* 69: 1009–1016.

24. E. R. Ingham and W. G. Thies, 1996. "Responses of soil foodweb organisms in the first year following clearcutting and application of chloropicrin to control laminated root rot," *Applied Soil Ecology* 3: 35–47.

25. A. S. Moffat, 1993. "Clearcutting's Soil Effects," *Science* 261: 1116.

26. Perry et al., 1990.

27. D. A. Perry, M. P. Amaranthus, J. G. Borchers, S. L. Borchers, and R. E. Brainerd, 1989. "Bootstrapping in ecosystems," *BioScience* 39: 230–237.

28. Perry et al., 1990.

29. E.-D. Schulze, F. A. Bazzaz, K. Nadelhoffer, T. Koike, and S. Takatsuki, 1996. "Biodiversity and ecosystem function of temperate deciduous broad-leaved forests," in H. A. Mooney, J. H. Cushman, E. Medina, O. E. Sala, and E.-D. Schulze (eds.), *Functional Roles of Biodiversity: A Global Perspective,* John Wiley & Sons, Chichester, U.K., pp. 71–98.

30. E.-D. Schulze and P. Gerstberger, 1993. "Functional aspects of landscape di-

versity: A Bavarian example," in E.-D. Schulze and H. A. Mooney (eds.), *Biodiversity and Ecosystem Function,* Springer-Verlag, Berlin, pp. 453–466.

31. R. R. Northup, Z. Yu, R. A. Dahlgren, and K. A. Vogt, 1995. "Polyphenol control of nitrogen release from pine litter," *Nature* 377: 227–229.

32. F. S. Chapin III, 1995. "New cog in the nitrogen cycle," *Nature* 377: 199–200.

33. Schulze et al., 1996.

34. A. R. Wallace, 1889. *Travels on the Amazon and Rio Negro,* Ward, Lock and Co., London, reprinted by Dover, New York.

35. Daily, 1995.

36. R. A. Houghton, 1994. "The worldwide extent of land-use change," *BioScience* 44: 305–313.

37. W. J. Junk and K. Furch, 1985. "The physical and chemical properties of Amazonian waters and their relationships with the biota," in G. T. Prance and T. E. Lovejoy (eds.), *Amazonia: Key Environments Series,* Pergamon Press, Oxford.

38. J. J. Ewel, 1986. "Designing agricultural ecosystems for the humid tropics," *Annual Review of Ecology and Systematics* 17: 245–271.

39. J. J. Ewel, M. J. Mazzarino, and C. W. Berish, 1991. "Tropical soil fertility changes under monocultures and successional communities of different structure," *Ecological Applications* 1(3): 289–302.

40. D. F. Waterhouse, 1974. "The biological control of dung," *Scientific American* 230 (4): 100–109.

41. R. J. Hobbs, D. A. Saunders, L. A. Lobry de Bruyn, and A. R. Main, 1993. "Changes in biota," in R. J. Hobbs and D. A. Saunders (eds.), *Reintegrating Fragmented Landscapes: Towards Sustainable Production and Nature Conservation,* Springer-Verlag, New York, pp. 65–106.

42. R. D. Hughes, M. Tyndale-Biscoe, and J. Walker, 1978. "Effects of introduced dung beetles on the breeding and abundance of the Australian bushfly," *Bulletin of Entomological Research* 68: 361–372.

43. K. E. Lee, 1991. "The diversity of soil organisms," in D. L. Hawksworth (ed.), *The Biodiversity of Microorganisms and Invertebrates: Its Role in Sustainable Agriculture,* CAB International, Wallingford, U.K., pp. 73–87.

44. P. Lavelle and B. Pashanasi, 1989. "Soil macrofauna and land management in Peruvian Amazonia (Yurimaguas, Loreto)," *Pedobiologia* 33: 283–291.

45. M. J. Swift and J. M. Anderson, 1993. "Biodiversity and ecosystem function in agricultural systems," in E.-D. Schulze and H. A. Mooney (eds.), *Biodiversity and Ecosystem Function,* Springer-Verlag, Berlin, pp. 15–41.

46. K. E. Lee, 1985. *Earthworms: Their Ecology and Relationships with Soils and Land Use,* Academic Press (Australia), Sydney.

47. F. A. Bazzaz, 1990. "The response of natural ecosystems to the rising CO_2 levels," *Annual Review of Ecology and Systematics* 21:167–196.

48. F. A. Bazzaz and E. D. Fajer, 1992. "Plant life in a CO_2-rich world," *Scientific American* 266 (1): 68–74.

49. P. M. Vitousek, 1994. "Beyond global warming: Ecology and global change," *Ecology* 75: 1861–1876.

50. E. Arnolds, 1991. "Decline of ectomycorrhizal fungi in Europe," *Agriculture, Ecosystems and Environment* 35: 209–244.

51. F. Berendse, 1993. "Ecosystem stability, competition, and nutrient cycling," in E.-D. Schulze and H. A. Mooney (eds.), *Biodiversity and Ecosystem Function,* Springer-Verlag, Berlin, pp. 409–431.

52. J. D. Aber, K. J. Nadelhoffer, P. Steudler, and J. M. Melillo, 1989. "Nitrogen saturation in Northern forest ecosystems," *BioScience* 39: 378–386.

53. K. Rosen, P. Gundersen, L. Tegnhammar, M. Johansson, and T. Frogner, 1992. "Nitrogen enrichment of Nordic forest ecosystems," *Ambio* 21: 364–368.

54. W. Durka, E.-D. Schulze, G. Gebauer, and S. Voerkelius, 1994. "Effects of forest decline on uptake and leaching of deposited nitrate determined from 15-N and 18-O measurements," *Nature* 372: 765–767.

55. L. O. Hedin, 1994. "Stable isotopes, unstable forest," *Nature* 372: 725–726.

Chapter Six

1. E.-D. Schulze, F. A. Bazzaz, K. Nadelhoffer, T. Koike, and S. Takatsuki, 1996. "Biodiversity and ecosystem function of temperate deciduous broad-leaved forests," in H. A. Mooney, J. H. Cushman, E. Medina, O. E. Sala, and E.-D. Schulze (eds.), *Functional Roles of Biodiversity: A Global Perspective,* John Wiley & Sons, Chichester, U.K., pp. 71–98.

2. P. M. Vitousek, P. R. Ehrlich, A. H. Ehrlich, and P. A. Matson, 1986. "Human appropriation of the products of photosynthesis," *BioScience* 36: 368–373.

3. R. A. Houghton, 1994. "The worldwide extent of land-use change," *BioScience* 44: 305–313.

4. The World Conservation Union, United Nations Environment Program, and World Wide Fund for Nature, 1991. *Caring for the Earth: A Strategy for Sustainable Living,* Gland, Switzerland.

5. D. M. Gates, 1985. *Energy and Ecology,* Sinauer Associates, Sunderland, Massachusetts.

6. C. J. Krebs, 1994. *Ecology,* 4th ed., HarperCollins College Publishers, New York.

7. M. G. Paoletti, D. Pimentel, B. R. Stinner, and D. Stinner, 1992. "Agroecosystem biodiversity: Matching production and conservation biology," *Agriculture, Ecosystems and Environment* 40: 3–23.

8. M. J. Swift, J. Vandermeer, P. S. Ramakrishnan, J. M. Anderson, C. K. Ong, and B. A. Hawkins, 1996. "Biodiversity and agroecosystem function," in H. A. Mooney, J. H. Cushman, E. Medina, O. E. Sala, and E.-D. Schulze (eds.), *Functional Roles of Biodiversity: A Global Perspective,* John Wiley & Sons, Chichester, U.K., pp. 261–298.

9. G. Gardner, 1996. "Preserving agricultural resources," in L. R. Brown et al. (eds.), *State of the World,* WorldWatch Institute/W.W. Norton & Co., New York, pp. 78–94.

10. M. W. Rosegrant and P. L. Pingall, 1991. "Sustaining rice productivity growth in Asia: A policy perspective," *IRRI Social Science Division Papers,* No. 91-01, International Rice Research Institute, Manila, Philippines.

11. J. Salick and L. C. Merrick, 1990. "Use and maintenance of genetic resources: Crops and their wild relatives," in C. R. Carroll, J. H. Vandermeer, and P. Rosset (eds.), *Agroecology,* McGraw-Hill, New York, pp. 517–548.

12. P. Raeburn, 1995. *The Last Harvest: The Genetic Gamble That Threatens to Destroy American Agriculture,* Simon & Schuster, New York.

13. R. Barbault and P. Lasserre, 1994. "Biodiversity: Nature in the balance," *UNESCO Sources* 60, July/August, p. 3.

14. Raeburn, 1995.

15. B. R. Trenbath, 1974. "Biomass productivity of mixtures," *Advances in Agronomy* 26: 177–210.

16. D. Pimentel et al., 1992. "Conserving biological diversity in agricultural/forestry systems," *BioScience* 42: 354–362.

17. World Conservation Union et al., 1991.

18. C. A. Francis, 1989. "Biological efficiencies in multiple-cropping systems," *Advances in Agronomy* 42: 1–42.

19. M. R. Rao and R. W. Willey, 1980. "Evaluation of yield stability in intercropping: Studies on sorghum/pigeonpea," *Experimental Agriculture* 16: 105–116.

20. Francis, 1989.

21. M. A. Altieri, 1991. "Increasing biodiversity to improve insect pest management in agro-ecosystems," in D. L. Hawksworth, *The Biodiversity of Microorganisms and Invertebrates: Its Role in Sustainable Agriculture,* CAB International, Wallingford, U.K., pp. 165–182.

22. Swift et al. 1996.

23. J. P. Haggar and J. J. Ewel, in press. "Primary productivity and resource partitioning in model tropical ecosystems," *Ecology.*

24. H. A. Mooney, J. Lubchenco, R. Dirzo, and O. E. Sala (eds.), 1995. "Biodiversity and Ecosystem Functioning," in United Nations Environment Program, *Global Biodiversity Assessment,* Cambridge University Press, Cambridge, pp. 275–452.

25. S. Naeem, L. J. Thompson, S. P. Lawler, J. H. Lawton, and R. M Woodfin, 1994. "Declining biodiversity can alter the performance of ecosystems," *Nature* 368: 734–737.

26. D. Tilman, D. Wedin, and J. Knops, 1996. "Productivity and sustainability influenced by biodiversity in grassland ecosystems," *Nature* 379: 718–720.

27. D. Tilman and J. A. Downing, 1994. "Biodiversity and stability in grasslands," *Nature* 367: 363–365.

28. T. J. Givnish, 1994. "Does diversity beget stability?" *Nature* 371: 113–114.

29. F. Berendse, 1993. "Ecosystem stability, competition, and nutrient cycling," in E.-D. Schulze and H. A. Mooney (eds.), *Biodiversity and Ecosystem Function,* Springer-Verlag, Berlin, pp. 409–431.

30. R. Aerts and F. Berendse, 1989. "Above-ground nutrient turnover and net primary production of an evergreen and a deciduous species in a heathland ecosystem," *Journal of Ecology* 77: 343–356.

31. D. Tilman, 1996. "Biodiversity: Population versus ecosystem stability," *Ecology* 77: 350–363.

32. F. I. Woodward, 1993. "How many species are required for a functional ecosystem?" in E.-D. Schulze and H. A. Mooney (eds.), *Biodiversity and Ecosystem Function,* Springer-Verlag, Berlin, pp. 271–291.

33. F. S. Chapin III and G. R. Shaver, 1985. "Individualistic growth response of tundra plant species to environmental manipulations in the field," *Ecology* 66: 564–576.

34. R. J. Hobbs and H. A. Mooney, 1991. "Effects of rainfall variability and gopher disturbance on serpentine annual grassland dynamics," *Ecology* 72: 59–68.

35. S. J. McNaughton, 1993. "Biodiversity and function of grazing ecosystems," in E.-D. Schulze and H. A. Mooney (eds.), *Biodiversity and Ecosystem Function,* Springer-Verlag, Berlin, pp. 361–383.

36. S. J. McNaughton, 1985. "Ecology of a grazing ecosystem: The Serengeti," *Ecological Monographs* 55: 259–294.

37. G. E. Rehfeldt, 1979. "Ecological adaptations in Douglas fir (*Pseudotsuga menziesii* var. *glauca*) populations: I. North Idaho and Northeast Washington," *Heredity* 43: 383–397.

38. F. S. Chapin III, 1980. "The mineral nutrition of wild plants," *Annual Review of Ecology and Systematics* 11: 233–260.

39. R. W. Pearcy and R. H. Robichaux, 1985. "Tropical and subtropical forests," in B. F. Chabot and H. A. Mooney (eds.), *Physiological Ecology of North American Plant Communities,* Chapman and Hall, New York, pp. 278–295.

40. P. M. Vitousek and L. R. Walker, 1989. "Biological invasion by *Myrica faya* in Hawaii: Plant demography, nitrogen fixation, ecosystem effects," *Ecological Monographs* 59: 247–265.

41. Mooney et al., 1995.

42. Ibid.

43. S. J. Wright, 1996. "Plant species diversity and ecosystem functioning in tropical forests," in G. H. Orians, R. Dirzo, and J. H. Cushman (eds.), *Biodiversity and Ecosystem Processes in Tropical Forests,* Springer-Verlag, Berlin, pp. 11–31.

Chapter Seven

1. S. A. Zimov, V. I. Chuprynin, A. P. Oreshko, F. S. Chapin III, M. C. Chapin, and J. F. Reynolds, 1995. "Effects of mammals on ecosystem change at the

Pleistocene-Holocene boundary," in F. S. Chapin III and C. Korner (eds.), *Arctic and Alpine Biodiversity: Patterns, Causes, and Ecosystem Consequences,* Springer-Verlag, Berlin, pp. 127–135.

2. R. Monastersky, 1994. "Earthmovers: Humans take their place alongside wind, water, and ice," *Science News* 146: 432–433.

3. R. J. Naiman, 1988. "Animal influences on ecosystem dynamics," *Bio-Science* 38: 750–752.

4. H. A. Mooney, J. Lubchenco, R. Dirzo, and O. E. Sala (eds.), 1995. "Biodiversity and Ecosystem Functioning," in United Nations Environment Program, *Global Biodiversity Assessment,* Cambridge University Press, Cambridge, pp. 275–452.

5. N. Owen-Smith, 1987. "Pleistocene extinctions: The pivotal role of megaherbivores," *Paleobiology* 13: 351–362.

6. D. H. M. Cumming, 1982. "The influence of large herbivores on savanna structure in Africa," in B. J. Huntley and B. H. Walker (eds.), *Ecology of Tropical Savannas,* Springer-Verlag, Berlin, pp. 217–245.

7. Owen-Smith, 1987.

8. N. Owen-Smith, 1989. "Megafaunal extinctions: The conservation message from 11,000 years B.P.," *Conservation Biology* 3: 405–412.

9. N. Owen-Smith, 1988. *Megaherbivores: The Influence of Very Large Body Size on Ecology,* Cambridge University Press, Cambridge.

10. H. T. Dublin, A. R. E. Sinclair, and J. McGlade, 1990. "Elephants and fire as causes of multiple stable states in the Serengeti-Mara woodlands," *Journal of Animal Ecology* 59: 1147–1164.

11. Owen-Smith, 1989.

12. R. M. Laws, 1981. "Experiences in the study of large mammals," in C. W. Fowler and T. D. Smith (eds.), *Dynamics of Large Mammal Populations,* John Wiley & Sons, New York, pp. 19–45.

13. David Western, personal communication.

14. Owen-Smith, 1989.

15. S. H. Jenkins and P. E. Busher, 1979. "*Castor canadensis,*" *Mammalian Species* 120: 1–8.

16. M. M. Remillard, G. K. Gruendling, and D. J. Bogucki, 1987. "Disturbance by beaver (*Castor canadensis* Kuhl) and increased landscape heterogeneity," in M. G. Turner (ed.), *Landscape Heterogeneity and Disturbance,* Springer-Verlag, New York, pp. 102–122.

17. J. Pastor, R. J. Naiman, and B. Dewey, 1987. "A hypothesis of the effects of moose and beaver foraging on soil nitrogen and carbon dynamics, Isle Royale," *Alces* 23: 107–123.

18. J. Pastor, R. J. Naiman, B. Dewey, and P. McInnes, 1988. "Moose, microbes, and the boreal forest," *BioScience* 38: 770–777.

19. J. Pastor, D. Mladenoff, Y. Haila, J. Bryant, and S. Payette, 1996. "Biodiversity and ecosystem processes in boreal regions," in H. A. Mooney, J. H. Cushman, E. Medina, O. E. Sala, and E.-D. Schulze (eds.), *Functional Roles*

of Biodiversity: A Global Perspective, John Wiley & Sons, Chichester, U.K., pp. 33–70.

20. Zimov et al., 1995.

21. A. D. Whicker and J. K. Detling, 1988. "Ecological consequences of prairie dog disturbances," *BioScience* 38: 778–785.

22. R. J. Naiman, G. Pinay, C. A. Johnston, and J. Pastor, 1994. "Beaver influences on the long-term biogeochemical characteristics of boreal forest drainage networks," *Ecology* 75: 905–921.

23. N. Huntly and R. Inouye, 1988. "Pocket gophers in ecosystems: Patterns and mechanisms," *BioScience* 38: 786–793.

24. S. E. Tardiff and J. A. Stanford, in preparation. "The effect of grizzly bear digging on subalpine meadow ecosystems."

25. D. H. Knight, 1987. "Parasites, lightning, and the vegetation mosaic in wilderness landscapes," in M. G. Turner (ed.), *Landscape Heterogeneity and Disturbance,* Springer-Verlag, New York, pp. 59–83.

26. W. D. Billings, 1990. "*Bromus tectorum,* a biotic cause of ecosystem impoverishment in the Great Basin," in G. M. Woodwell (ed.), *The Earth in Transition: Patterns and Processes of Biotic Impoverishment,* Cambridge University Press, Cambridge, pp. 301–322.

27. R. N. Mack, 1986. "Alien plant invasion into the Intermountain West: A case history," in H. A. Mooney and J. A. Drake (eds.), *Ecology of Biological Invasions of North America and Hawaii,* Springer-Verlag, New York, pp. 191–213.

28. Billings, 1990.

29. R. Devine, 1993. "The cheatgrass problem," *The Atlantic Monthly,* May, pp. 40–48.

30. P. H. Zedler, C. R. Gautier, and G. S. McMaster, 1983. "Vegetation change in response to extreme events: The effect of a short interval between fires in California chaparral and coastal scrub," *Ecology* 64: 809–818.

31. F. Hughes, P. M. Vitousek, and T. Tunison, 1991. "Alien grass invasion and fire in the seasonal submontane zone of Hawaii," *Ecology* 72: 743–746.

32. C. R. Carroll, 1990. "The interface between natural areas and agroecosystems," in C. R. Carroll, J. H. Vandermeer, and P. Rousset (eds.), *Agroecology,* McGraw-Hill Publishing Co., New York, pp. 365–383.

33. U.S. Congress, Office of Technology Assessment, 1993. *Harmful Non-indigenous Species in the United States,* OTA-F-565, U.S. Government Printing Office, Washington, D.C.

34. J. J. Ewel, 1986. "Invasibility: Lessons from south Florida," in H. A. Mooney and J. A. Drake (eds.), *Ecology of Biological Invasions of North America and Hawaii,* Springer-Verlag, New York, pp. 214–230.

35. U.S. Congress, Office of Technology Assessment, 1993.

36. E. Garcia-Moya and C. M. McKell, 1970, "Contribution of shrubs to the nitrogen economy of a desert-wash plant community," *Ecology* 51: 81–88.

37. M. H. Friedel, B. D. Foran, and D. M. Stafford Smith, 1990. "Where the

creeks run dry or ten feet high: Pastoral management in arid Australia," *Proceedings of the Ecological Society of Australia* 16: 185–194.

38. L. P. White, 1971. "Vegetation stripes on sheet wash surfaces," *Journal of Ecology* 59: 615–622.

39. D. J. Tongway and J. A. Ludwig, 1990. "Vegetation and soil patterning in semi-arid mulga lands of Eastern Australia," *Australian Journal of Ecology* 15: 23–34.

40. P. G. Risser, 1995. "The status of the science examining ecotones," *BioScience* 45: 318–325.

41. The World Conservation Union, United Nations Environment Program, and World Wide Fund for Nature, 1991. *Caring for the Earth: A Strategy for Sustainable Living,* Gland, Switzerland.

42. T. J. Done, J. C. Ogden, and W. J. Wiebe, 1996. "Biodiversity and ecosystem function of coral reefs," in H. A. Mooney, J. H. Cushman, E. Medina, O. E. Sala, and E.-D. Schulze (eds.), *Functional Roles of Biodiversity: A Global Perspective,* John Wiley & Sons, Chichester, U.K., pp. 393–429.

43. R. H. Richmond, 1993. "Coral reefs: present problems and future concerns resulting from anthropogenic disturbance," *American Zoologist* 33: 524–536.

Chapter Eight

1. J. Shukla, C. Nobre, and P. Sellers, 1990. "Amazon deforestation and climate change," *Science* 247: 1322–1325.

2. G. P. Marsh, 1874. *The Earth as Modified by Human Action,* Schreibner, Armstrong & Co., New York (reprint edition 1970, Arno Press Inc.), p. 196.

3. R. A. Anthes, 1984. "Enhancement of convective precipitation by mesoscale variations in vegetative covering in semiarid regions," *Journal of Climate and Applied Meteorology* 23: 541–554.

4. K. Thompson, 1980. "Forests and climate change in America: Some early views," *Climatic Change* 3: 47–64.

5. Marsh, 1874, p. 195.

6. C. W. Thornthwaite, 1956. "Modification of rural microclimates," in W. L. Thomas Jr. (ed.), *Man's Role in Changing the Face of the Earth,* Vol. 2, University of Chicago Press, Chicago, pp. 567–583.

7. Y. Mintz, 1984. "The sensitivity of numerically simulated climates to land-surface boundary conditions" in J. T. Houghton (ed.), *The Global Climate,* Cambridge University Press, Cambridge, pp. 79–105.

8. Anthes, 1984.

9. C. Sagan, O. B. Toon, and J. B. Pollack, 1979. "Anthropogenic albedo changes and the earth's climate," *Science* 206: 1363–1368.

10. J. Shukla and Y. Mintz, 1982. "Influence of land-surface evapotranspiration on the earth's climate," *Science* 215: 1498–1500.

11. Shukla et al., 1990.
12. E. Salati, 1985. "The climatology and hydrology of Amazonia," in G. T. Prance and T. E. Lovejoy (eds.), *Amazonia: Key Environments Series,* Pergamon Press, Oxford, pp. 18–48.
13. Shukla et al., 1990.
14. Shukla and Mintz, 1982.
15. G. B. Bonan, D. Pollard, and S. L. Thompson, 1992. "Effects of boreal forest vegetation on global climate," *Nature* 359: 716–718.
16. Y. Baskin, 1994. "The greening of global climate models," *Earth,* March, pp. 26–32.
17. FAUNMAP Working Group, 1996. "Spatial response of mammals to late Quaternary environmental fluctuations," *Science* 272: 1601–1606.
18. J. A. Foley, J. E. Kutzback, M. T. Coe, and S. Levis, 1994. "Feedback between climate and boreal forests during the Holocene epoch," *Nature* 371: 52–54.
19. P. J. Sellers et. al., 1996. "Comparison of radiative and physiological effects of doubled atmospheric CO_2 on climate," *Science* 271: 1402–1406.
20. T. J. Lyons, P. Schwerdtfeger, J. M. Hacker, I. J. Foster, R. C. G. Smith, and H. Xinmei, 1993. "Land–atmosphere interaction in a semiarid region: The bunny fence experiment," *Bulletin of the American Meteorological Society* 74: 1327–1334.
21. J. Charney, P. H. Stone, and W. J. Quirk, 1975. "Drought in the Sahara: A biogeophysical feedback mechanism," *Science* 187: 434–435.
22. J. Otterman, 1974. "Baring high-albedo soils by overgrazing: A hypothesized desertification mechanism," *Science* 186: 531–533.
23. W. H. Schlesinger, J. F. Reynolds, G. L. Cunningham, L. F. Huenneke, W. M. Jarrell, R. A. Virginia, and W. G. Whitford, 1990. "Biological feedbacks in global desertification," *Science* 247: 1043–1048.
24. R. A. Rice and J. Vandermeer, 1990. "Climate and the geography of agriculture," in C. R. Carroll, J. H. Vandermeer, and P. Rosset (eds.), *Agroecology,* McGraw-Hill Publishing Co., New York, pp. 21–63.
25. National Academy of Sciences, National Academy of Engineering, and Institute of Medicine, 1992. *Policy Implications of Greenhouse Warming: Mitigation, Adaptation and the Science Base,* National Academy Press, Washington, D.C.
26. R. A. Houghton, 1994. "The worldwide extent of land-use change," *BioScience* 44: 305–313.
27. R. A. Kerr, 1995. "It's official: First glimmer of greenhouse warming seen," *Science* 270: 1565–1567.
28. VEMAP Members, 1995. "Vegetation/ecosystem modeling and analysis project: Comparing biogeography and biogeochemistry models in a continental-scale study of terrestrial ecosystem responses to climate change and CO_2 doubling," *Global Biogeochemical Cycles* 9:407–437.
29. R. A. Kerr, 1995. "Greenhouse report foresees growing global stress," *Science* 270: 731.

30. R. H. Waring and W. H. Schlesinger, 1985. *Forest Ecosystems: Concepts and Management,* Academic Press, Orlando, Florida.
31. R. K. Dixon, S. Brown, R. A. Houghton, A. M. Solomon, M. C. Trexler, and J. Wisniewski, 1994. "Carbon pools and flux of global forest ecosystems," *Science* 263: 185–190.
32. H. A. Mooney, B. G. Drake, R. J. Luxmoore, W. C. Oechel, and L. F. Pitelka, 1991. "Predicting ecosystem responses to elevated CO_2 concentrations," *BioScience* 41: 96–104.
33. F. A. Bazzaz and E. D. Fajer, 1992. "Plant life in a CO_2-rich world," *Scientific American* 266: 68–74.
34. National Academy of Sciences et al., 1992.
35. Y. Baskin, 1995. "Can iron supplementation make the equatorial Pacific bloom?" *BioScience* 45: 314–316.
36. R. Monastersky, 1995. "Iron versus the greenhouse," *Science News* 148: 220–222.
37. National Academy of Sciences et al., 1992.
38. G. Marland, 1988. *The Prospect of Solving the CO_2 Problem through Global Reforestation,* U.S. Department of Energy, Report DOE/NBB-0082, Oak Ridge National Laboratory, Oak Ridge, Tennessee.
39. Dixon et al., 1994.
40. A. Mosier, D. Schimel, D. Valentine, K. Bronson, and W. Parton, 1991. "Methane and nitrous oxide fluxes in native, fertilized and cultivated grasslands," *Nature* 350: 330–332.
41. H. A. Mooney, J. Lubchenco, R. Dirzo, and O. E. Sala (eds.), 1995. "Biodiversity and Ecosystem Functioning," in United Nations Environment Program, *Global Biodiversity Assessment,* Cambridge University Press, Cambridge, pp. 275–452.
42. R. L. Sass, 1995. "Mitigation of methane emissions from irrigated rice agriculture," *Global Change Newsletter, The International Geosphere-Biosphere Program* 22: 4–5.
43. H.-U. Neue, 1993. "Methane emissions from rice fields," *BioScience* 43: 466–474.
44. S. C. Whalen and W. S. Reeburgh, 1990. "Consumption of atmospheric methane by tundra soils," *Nature* 346: 160–162.
45. P. A. Steudler, R. D. Bowden, J. M. Melillo, and J. D.Aber, 1989. "Influence of nitrogen fertilization on methane uptake in temperate forest soils," *Nature* 341: 314–316.
46. Mosier et al., 1991.
47. B. W. Hutsch, C. P. Webster, and D. S. Powlson, 1993. "Long-term effects of nitrogen fertilization on methane oxidation in soil of the Broadbalk wheat experiment," *Soil Biology and Biochemistry* 25: 1307–1315.
48. G. J. Whiting and J. P. Chanton, 1993. "Primary production control of methane emission from wetlands," *Nature* 364: 794–795.
49. J. W. H. Dacey, B. G. Drake, and M. J. Klug, 1994. "Stimulation of methane

emission by carbon dioxide enrichment of marsh vegetation," *Nature* 370: 47–49.

50. Mosier et al., 1991.

51. M. Keller, T. J. Goreau, S. C. Wofsy, W. A. Kaplan, and M. B. McElroy, 1983. "Production of nitrous oxide and consumption of methane by forest soils," *Geophysical Research Letter* 10: 1156–1159.

52. M. Keller, E. Veldkamp, A. M. Weitz, and W. A. Reiners, 1993. "Effect of pasture age on soil trace-gas emissions from a deforested area of Costa Rica," *Nature* 365: 244–246.

53. J. W. H. Dacey and S. G. Wakeham, 1986. "Oceanic dimethylsulfide: Production during zooplankton grazing on phytoplankton," *Science* 233: 1314–1316.

54. R. J. Charlson, J. E. Lovelock, M. O. Andreae, and S. G. Warren, 1987. "Oceanic phytoplankton, atmospheric sulphur, cloud albedo and climate," *Nature* 326: 655–661.

55. J. E. Lovelock, 1979. *Gaia: A New Look at Life on Earth,* Oxford University Press, Oxford.

56. A. J. Watson, 1991. "Gaia," *New Scientist* 131: 1–4.

57. J. E. Lovelock and L. R. Kump, 1994. "Failure of climate regulation in a geophysiological model," *Nature* 369: 732–735.

58. Mooney et al., 1995.

Chapter Nine

1. M. Nelson, T. L. Burgess, A. Alling, N. Alvarez-Romo, W. F. Dempster, R. L. Walford, and J. P. Allen, 1993. "Using a closed ecological system to study earth's biosphere," *BioScience* 43: 225–236.

2. T. Appenzeller, 1994. "Biosphere 2 makes a new bid for scientific credibility," *Science* 263: 1368–1369.

3. Tony L. Burgess, personal communication.

4. W. J. Mitsch, 1996. "Ecological engineering: A new paradigm for engineers and ecologists," in P. C. Schulze (ed.), *Engineering within Ecological Constraints,* National Academy Press, Washington, D.C., pp. 111–128.

5. E. C. Wolf, 1987. "On the brink of extinction: Conserving the diversity of life," *Worldwatch Paper* 78.

6. L. Roberts, 1993. "Wetlands trading is a loser's game, say ecologists," *Science* 260: 1890–1892.

7. J. B. Zedler, 1993. "Restoring biodiversity to coastal salt marshes," in J. E. Keeley (ed.), *Interface between Ecology and Land Development in California,* Southern California Academy of Sciences, Los Angeles, pp. 253–257.

8. J. B. Zedler, 1993. "Canopy architecture of natural and planted cordgrass marshes: Selecting habitat evaluation criteria," *Ecological Applications* 3: 123–138.

9. K. D. Gibson, J. B. Zedler, and R. Langis, 1994. "Limited response of cord-grass *(Spartina foliosa)* to soil amendments in a constructed marsh," *Ecological Applications* 4: 757–767.

10. J. B. Zedler, 1988. "Restoring diversity in salt marshes: Can we do it?" in E. O. Wilson (ed.), *Biodiversity,* National Academy Press, Washington, D.C., pp. 317–325.

11. National Research Council, 1992. *Restoration of Aquatic Ecosystems: Science, Technology, and Public Policy,* National Academy Press, Washington, D.C.

12. F. H. Bormann, 1976. "An inseparable linkage: Conservation of natural ecosystems and conservation of fossil fuel," *BioScience* 26: 754–760.

13. E. P. Odum and H. T. Odum, 1972. "Natural areas as necessary components of man's total environment," *Transactions of the North American Wildlife and Natural Resources Conference* 37: 178–189.

14. B. Norton, 1988. "Commodity, amenity, and morality: The limits of quantification in valuing biodiversity," in E. O. Wilson (ed.), *Biodiversity,* National Academy Press, Washington, D.C., pp. 200–205.

15. T. E. Lovejoy, 1995. "Will expectedly the top blow off?" *BioScience,* Science and Biodiversity Policy Supplement: S-3-6.

16. D. Pimentel et al., 1992. "Conserving biological diversity in agricultural/forestry systems," *BioScience* 42: 354–362.

17. R. Goodland and G. Ledec, 1989. "Wildlands: Balancing conversion with conservation in World Bank projects," *Environment* 31: 6–11, 27–35.

18. Pimentel et al., 1992.

19. U. Narain and A. Fisher, 1994. "Modeling the value of biodiversity using a production function approach," in C. Perrings et al. (eds.), *Biodiversity Conservation: Policy Issues and Options,* Kluwer Academic Publishers, Dordrecht.

20. P. Raeburn, 1995. *Last Harvest,* Simon & Schuster, New York.

21. E. B. Barbier, J. C. Burgess, and C. Folke, 1994. *Paradise Lost? The Ecological Economics of Biodiversity,* Earthscan Publications Ltd., London.

22. B. W. van Wilgen, R. M. Cowling, and C. J. Burgers, 1996. "Valuation of ecosystem services: A case study from South African fynbos ecosystems," *BioScience* 46: 184–189.

23. S. L. Pimm, G. J. Russell, J. L. Gittleman, and T. M. Brooks, 1995. "The future of biodiversity," *Science* 269: 347–350.

24. Norton, 1988.

25. D. Western, 1989. "Why manage nature?" in D. Western and M. C. Pearl (eds.), *Conservation for the Twenty-First Century,* Oxford University Press, New York, pp. 133–137.

26. Ocean Voice International, 1995. "Status of the world ocean and its biodiversity," *Sea Wind* 9 (4): 1–72.

27. N. Myers, 1993. "Noah's Choice," *Earthwatch,* July/August, pp. 20–21.

Index